Chapter 1. People Living on an Edge

This report is about people living on an edge, an unstable edge of sand arcing into the North Atlantic at the tip of Cape Cod, Massachusetts. They are dune dwellers, self-reliant mavericks choosing to live seasonally in fragile shacks perched on a ridge of barrier dunes above the sea. Here in winter, the empty shacks are exposed to battering northeast storms from the open ocean. From spring through fall when the shacks are used, they are bathed with reflected light from unrestricted horizons. The shacks float on what some dune dwellers call a "liquid earth," dunes that move, an unstable landscape of sand, wet berry bogs, thorny heaths, and stunted patches of pitch pine and oak. For generations the special

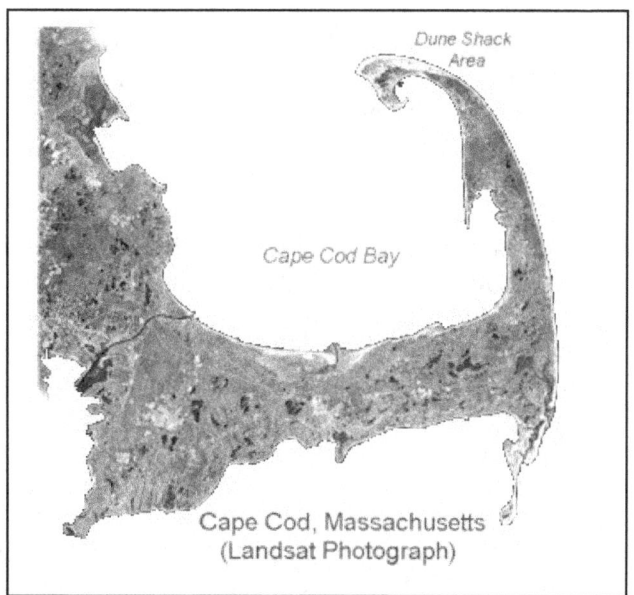

Cape Cod, Massachusetts
(Landsat Photograph)

qualities at the cape's farthest end have drawn fishers, artists, writers, and authentic eccentrics, forming into distinctive communities and establishing roots. It's a wondrous edge, the dune dwellers assert, offering solitude, new vistas, and creative energy. It's why they've lived there in simple shacks for generations, to be intimately connected to it, purposively exposed on the outer edge of the continent.

This report is ethnography, an anthropological picture of the traditions and cultures of the dune dwellers on the outside edge of the tip of Cape Cod. My goal in the report is to describe the seasonal settlement of the dune dwellers, to document their traditional cultural practices, beliefs, customs, and histories that are linked to the shacks and dunes. Over the years, the cape's end has drawn other peoples as well, such as surf fishers, dune buggy enthusiasts, and hoards of summer tourists. But this report does not focus on them. It focuses on long-term residents of the lower cape who have lived on the dunes in shacks. To understand the traditional culture of the dunes, I sought out experts, those people with a continuity of memory and direct experience connected to the shacks, dune dwellers who knew their own histories and practices.

There have never been many dune shacks. In 2004, the year of this study, they numbered just eighteen or nineteen shacks, depending on how you counted them. Nor have there been many dune dwellers. At present, about 250 people formed a core, with connections to perhaps another 1,100 to 1,700 users. The barrier dunes with shacks were not extensive either, only about three miles of shoreline, a viewshed of about 1,500 acres. But I discovered a wealth of cultural practices connected to this small area. And the people living on the dunes in shacks were eager to share them with me.

On the Edge of a Sandy Hook

Cape Cod is a long peninsula commonly described as a flexed arm stuck out into the sea, pointing northward. It's the remnant from the last great ice age, according to geologists. The

Wisconsin ice sheet shoved into the North Atlantic, riding on a deep moraine of rocky debris scoured from the continent. When the ice sheet retreated about 15,000 years ago, it left its rocky moorings, the terminal moraine, protruding into the ocean, a footing for what was to become the cape. Over the years woodlands covered the cape with pitch pine, oak, bearberry, and other plants. It became a homeland for Native Americans at least 3,000 years ago. When the Pilgrims arrived in the New World in 1620, they first landed at its northernmost bay a few miles from where the shacks sit today.

The tip of Cape Cod is not firm glacial moraine but unstable sand. Over the post-glacier millennia, ocean currents and storms have carried sand northward and southward along the outer rim of the cape, depositing, removing, and re-depositing it. A long beach formed on the outer edge, the great Outer Beach, miles of sandy shoreline with occasional spits protecting rich tidal wetlands. At the north end, the sand swept into deep water. Prevailing currents and storm action dropped the gritty loads, building up an end to the cape made almost completely of sand. The sweep of forces formed a "recurved hook" encircling a small body of salt water, a common formation at the ends of sand spits, resembling the crook of a gentleman's cane. However, this sandy hook was distinctive because it encircled a deep-water bay. The association of deep water and sand dunes is a rare formation. This sheltered bay became Provincetown Harbor, anchoring human settlement at the cape's tip. Mariners stopping at the port founded Provincetown, a settlement on sand that developed into a small but significant fishing town by the 18th century, harboring large fleets of whalers, fishers, and sea merchants. Provincetown sits on the inside of the sandy hook. It has always been a narrow strip of buildings hugging the beach. Behind it are ridges of sand dunes.

The cape's northern end is relatively narrow, less than a mile across. Its terrain is shaped into several curved dune ridges with valleys in between, the highest dunes standing about 80 feet tall. The ridges and valleys are remnants of successive barrier dune ridges left inland as the outer beach expanded. The sheltered valleys between dune ridges are frequently damp, sometimes boggy, and filled with heath-like patches of cranberry, blueberry, and bearberry. Salt runs extend up the valley mouths where they touch the sea. The dune slopes are anchored with dune grass and small patches of woodlands, primarily stunted pitch pine and oak. The woodlands and bogs support small populations of deer, coyotes, hare, and turtles. Seasonally, migratory waterfowl use the ponds, and shorebirds like piping plovers and terns frequent the outer beach for breeding and feeding.

The outside of the hook features a short barrier dune and a beach, exposed to winter weather called "brutal" by dune shack residents. The outside of the hook is where the shacks are located. They are built on one side or the other of the barrier dunes, close to vistas looking out onto the ocean. Residents of Provincetown call this thin sandy strip simply "the Backshore," contrasting it with the hook's inner shore where the town sits.

Stretching out from the Backshore, the ocean covers shoals, ridges of sand just beneath the waves. These are the infamous Peaked Hill Bars, several shallow sand bars evidently named for a prominent dune once visible to mariners from the sea. Locally, the name is pronounced with two syllables, "peak-ed," like someone looking sickly. The Peaked Hill Bars were infamous for claiming ships, as were all the shoals along the great Outer Beach. The sands are filled with the bones of wrecks. During the 19th century, a string of lifesaving stations operated along the Outer Beach to assist the shipwrecked. As shown later in the report, some dune dwellers of the Backshore traced their roots to the lifesavers of the Peaked Hill Station.

But the Backshore is more than a geomorphic edge. It's a conceptual edge as well. It represents an outer edge of three traditions described later in this report, sets of ideas, beliefs, and values. These cultural traditions give meaning to the lives of contemporary dune dwellers who choose to live at this edge in shacks, exploring their potentials.

First, the Backshore is the farthest edge of what I will call "Old Provincetown." This is the conceptual center of modern Provincetown, an iconic town for a year-round population that traces roots to the Portuguese and Yankees who harvested the sea's bounty from its sandy hook. The Backshore provides an outer fringe for the expression of traditions connected to Old Provincetown, like foraging, refuge from the stresses of town life, and training children.

Second, the Backshore is the outermost edge of Provincetown's fine arts colony. Formed during the late 19th century, the colony is reputed to be America's oldest, a gathering of artists, writers, and performers that has significance for the history of fine arts. Its creative expression extended out into the dunes, nurtured within the shacks, finding creative expression in unfettered solitude and reflected light.

And third, for some dune dwellers, the Backshore represented an edge to society. It's where human society ended, abutted, and joined untamed Nature. Conceptually, it's that place where the social philosopher, Henry Thoreau, stood during the mid-19th century and proclaimed, "A man may stand there and put all America behind him." The practices of some contemporary dune dwellers were demonstrations of those ideas.

So geographically, the Backshore comprises a thin outer fringe of a sandy spit. And demographically, it contains a small seasonal settlement of rustic shacks. Yet in sociocultural terms, the activities grounded in this landscape appear much larger. The dune residents believe their ways of living are parts of something culturally rich and complex, a layering of old and living traditions. And at times, it seems that the activities of dune shack residents have influenced the course of cultural traditions far beyond the cape.

Study Background

The sponsor of this ethnographic research is the National Park Service, the federal agency with oversight of America's national parks and monuments. The National Park Service funded the research because of a strong interest in the dune shacks – the Park owns them. The Park Service acquired the dunes and the shacks after the creation of the Cape Cod National Seashore.

Congress established the Cape Cod National Seashore in 1961 in recognition of the national significance of the great Outer Beach. The Seashore encompassed the entire beach facing the Atlantic Ocean running the length of the peninsula, and also portions of the eastern bayshore. The Seashore acquired lands within the towns of Chatham, Orleans, Eastham, Wellfleet, Truro, and Provincetown. The Backshore and the Province Lands near Provincetown became a part of the Seashore. After the Seashore's formation, the Park acquired all but one of the dune shacks as federal properties. Previously, the shacks were privately owned. In most cases, the owners of the shacks differed from the owners of the land underneath the shacks. Most of the Backshore had been parceled up into narrow strips of land that were privately held (so-called "spaghetti strips"). One shack, the westernmost, sat on the Province Lands held by the Commonwealth of Massachusetts. Under federal ownership, some shacks were demolished. But eighteen shacks continued as dwellings through special reservations or leases with dune shack residents, and a nineteenth shack (the Malicoat-Lord shack) continued as a privately-owned cottage on private land. In 1989, after nomination by the State, the district was found eligible for listing on the National Register of Historic Places. An historic district was established, called the Dune Shacks of the Peaked Hill Bars Historic District. The designation recognized historic values associated with the group of shacks. The historic district encompassed approximately 1,500 acres of land within the viewshed of the shacks.

Interest in documenting the traditional culture of the dune shack district emerged from a proposal from Provincetown's municipal government. The town's selectmen proposed that federal management preserve the traditions and culture associated with the dune shack district. These were additional values of the historic district, they asserted. Traditions and cultural patterns linked to the dune shacks had not been systematically assessed. Given this, the Park Service saw the utility of doing so, the impetus for this research project. The project was designed to provide ethnographic information and analysis to augment existing knowledge about the historic significance of the district.

Shacks and Trails

The dune shacks are small, weathered, and rustic-looking. They range from one-room structures to multi-room cottages. They are like "old people you admire," said one dune shack resident, perched at the outer tip of Cape Cod. They "watched" the waters, year after year, "collecting energy," "collecting stories."

The dune shacks of the Peaked Hill Bars Historic District lie within the fringes of Provincetown and North Truro. The shacks' locations are shown in Maps No. 1 to 4, along with the major trails in the district. Most shacks are short hikes from Provincetown by footpaths, indicated on the vicinity map (Map No. 1). Many small trails extend into the dunes from Provincetown, crisscrossing the district. Some trails were substantial wagon roads at one time, but most have shrunk into modest footpaths and faint deer runs by lack of traffic, traces weaving among patches of bracken and wood.

Top: Tasha shack. Bottom: Schnell-Del Deo shack. (Wolfe, 2004)

At the time of this study, many dune dwellers routinely walked between their shacks and town. But dune dwellers also reached shacks with over-sand vehicles, typically four-wheel-drive cars with partially-deflated tires. Three jeep trails entered the dunes, each providing access to one of three shack groups. One jeep trail entered the dunes from Race Point near the Provincetown airport, accessing the western shack group (Map No. 2). A second jeep trail entered the dunes near the west end of Pilgrim Lake (East Harbor), accessing the central shack group (Map No. 3). A third jeep trail entered the dunes at the foot of a headland known as High Head, accessing one shack in what was once an eastern shack group (Map No. 4).

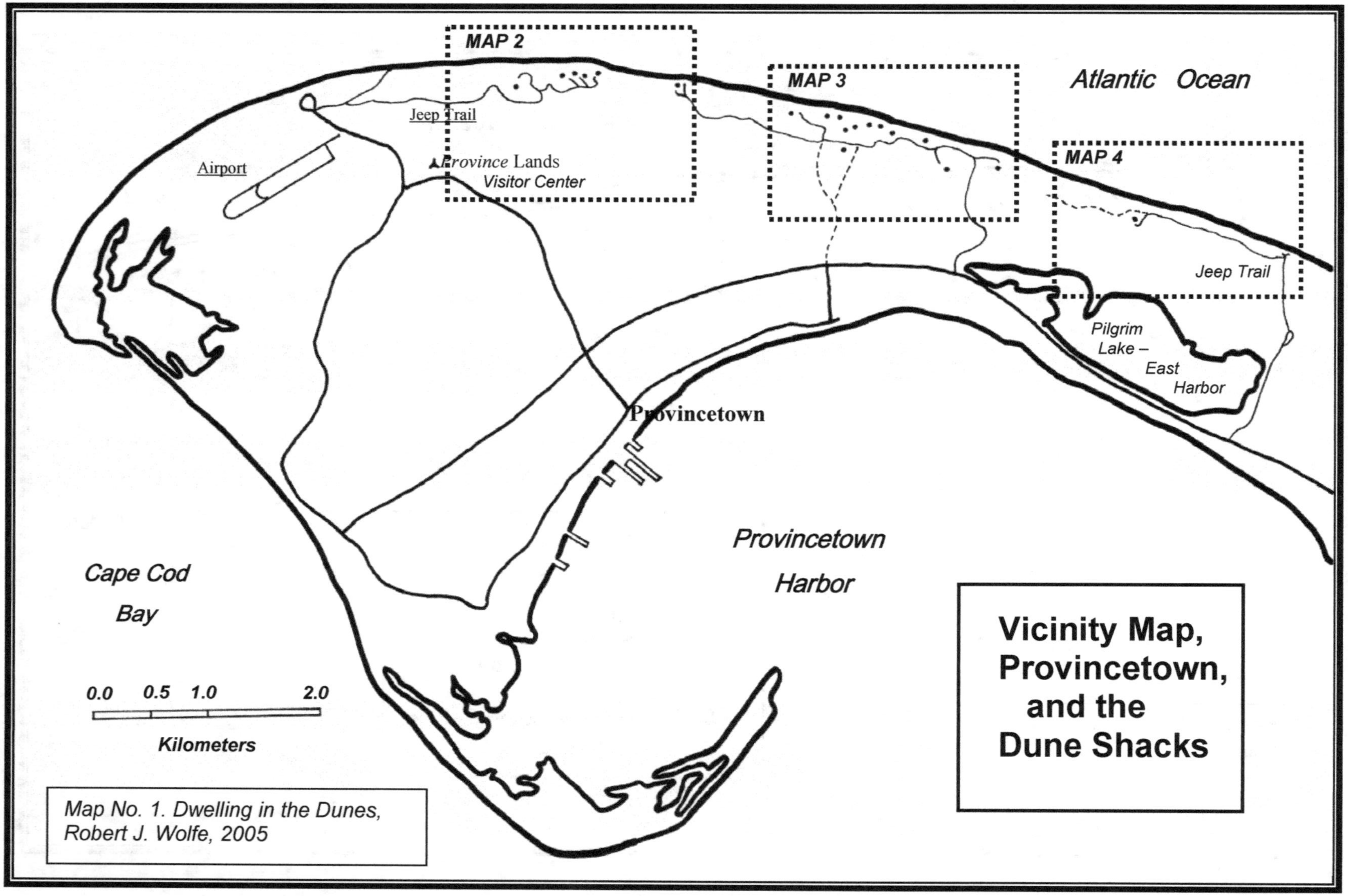

Vicinity Map, Provincetown, and the Dune Shacks

Atlantic Ocean

MAP 4

Jeep Trail

Pilgrim Lake – East Harbor

MAP 3

MAP 2

Jeep Trail

Province Lands Visitor Center

Airport

Provincetown

Provincetown Harbor

Cape Cod Bay

Kilometers

0.0   0.5   1.0   2.0

Map No. 1. Dwelling in the Dunes, Robert J. Wolfe, 2005

Chapter 1. People Living on an Edge

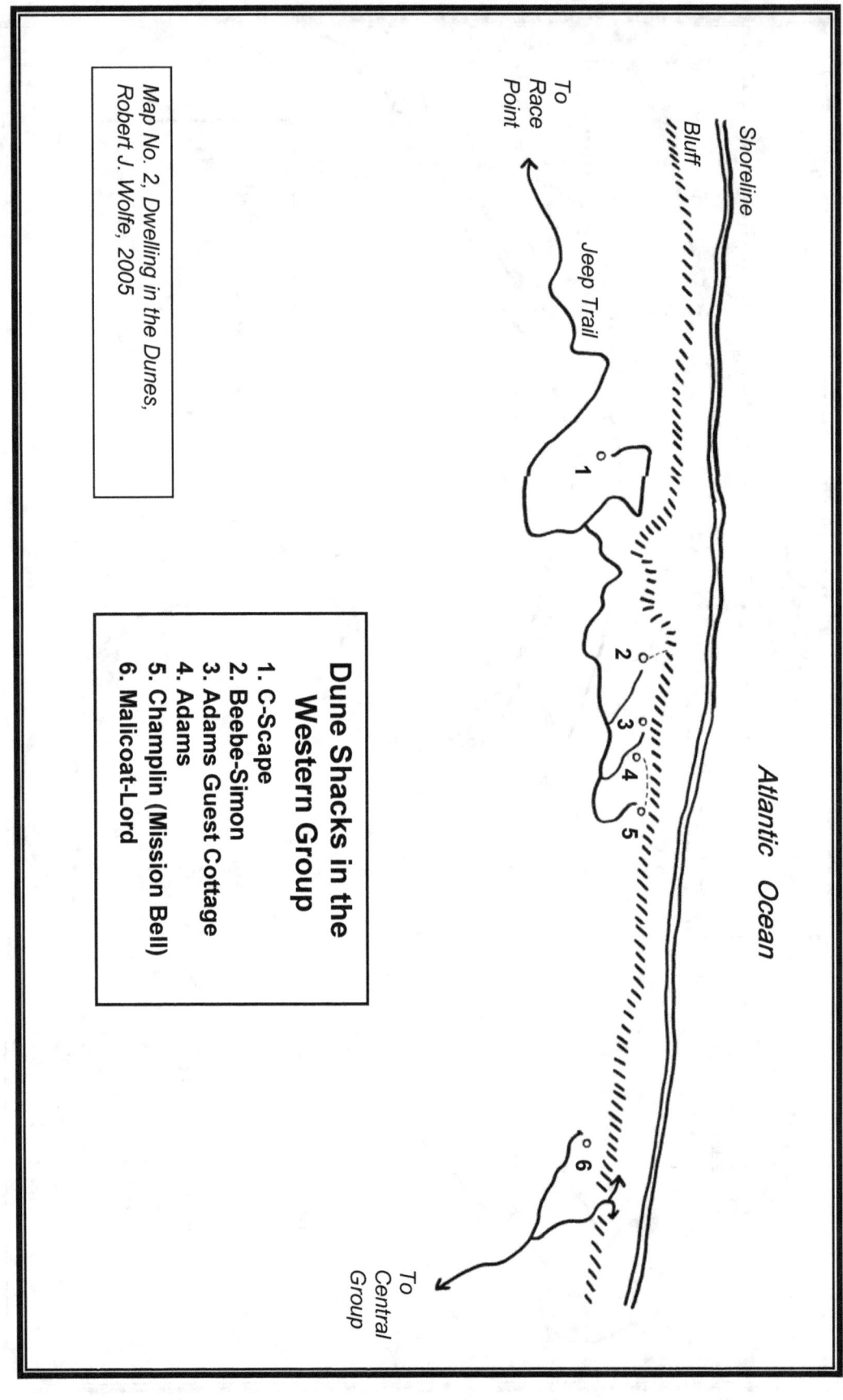

Atlantic Ocean

Shoreline

Bluff

To
Race
Point

Jeep Trail

**Dune Shacks in the
Western Group**

1. C-Scape
2. Beebe-Simon
3. Adams Guest Cottage
4. Adams
5. Champlin (Mission Bell)
6. Malicoat-Lord

To
Central
Group

Map No. 2, Dwelling in the Dunes,
Robert J. Wolfe, 2005

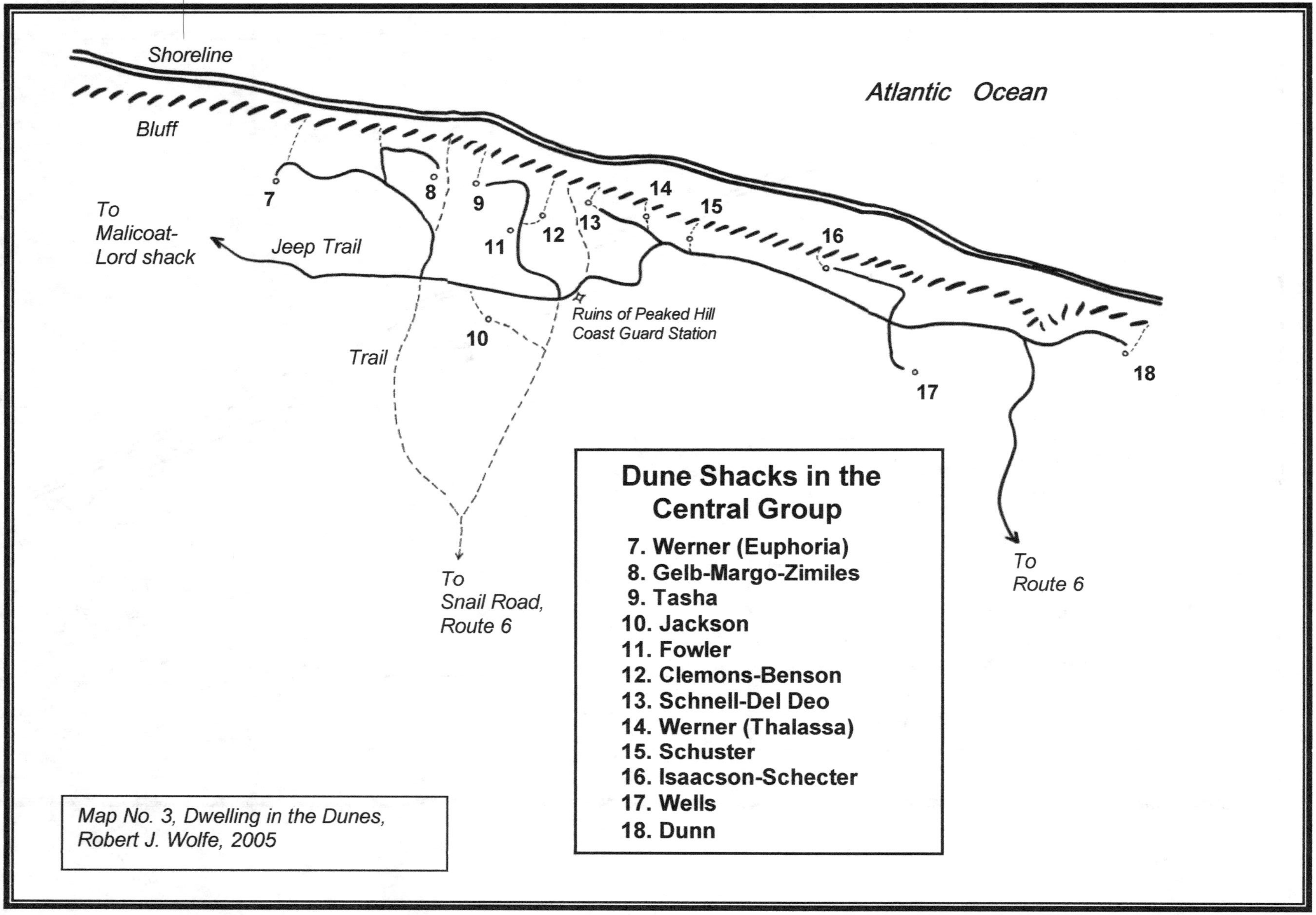

Atlantic Ocean

Shoreline
Bluff

To
Malicoat-
Lord shack

Jeep Trail

Trail

7

8

9

11

12

13

10

14

15

16

17

18

Ruins of Peaked Hill
Coast Guard Station

To
Snail Road,
Route 6

To
Route 6

Dune Shacks in the Central Group

7. Werner (Euphoria)
8. Gelb-Margo-Zimiles
9. Tasha
10. Jackson
11. Fowler
12. Clemons-Benson
13. Schnell-Del Deo
14. Werner (Thalassa)
15. Schuster
16. Isaacson-Schecter
17. Wells
18. Dunn

Map No. 3, Dwelling in the Dunes,
Robert J. Wolfe, 2005

Chapter 1. People Living on an Edge

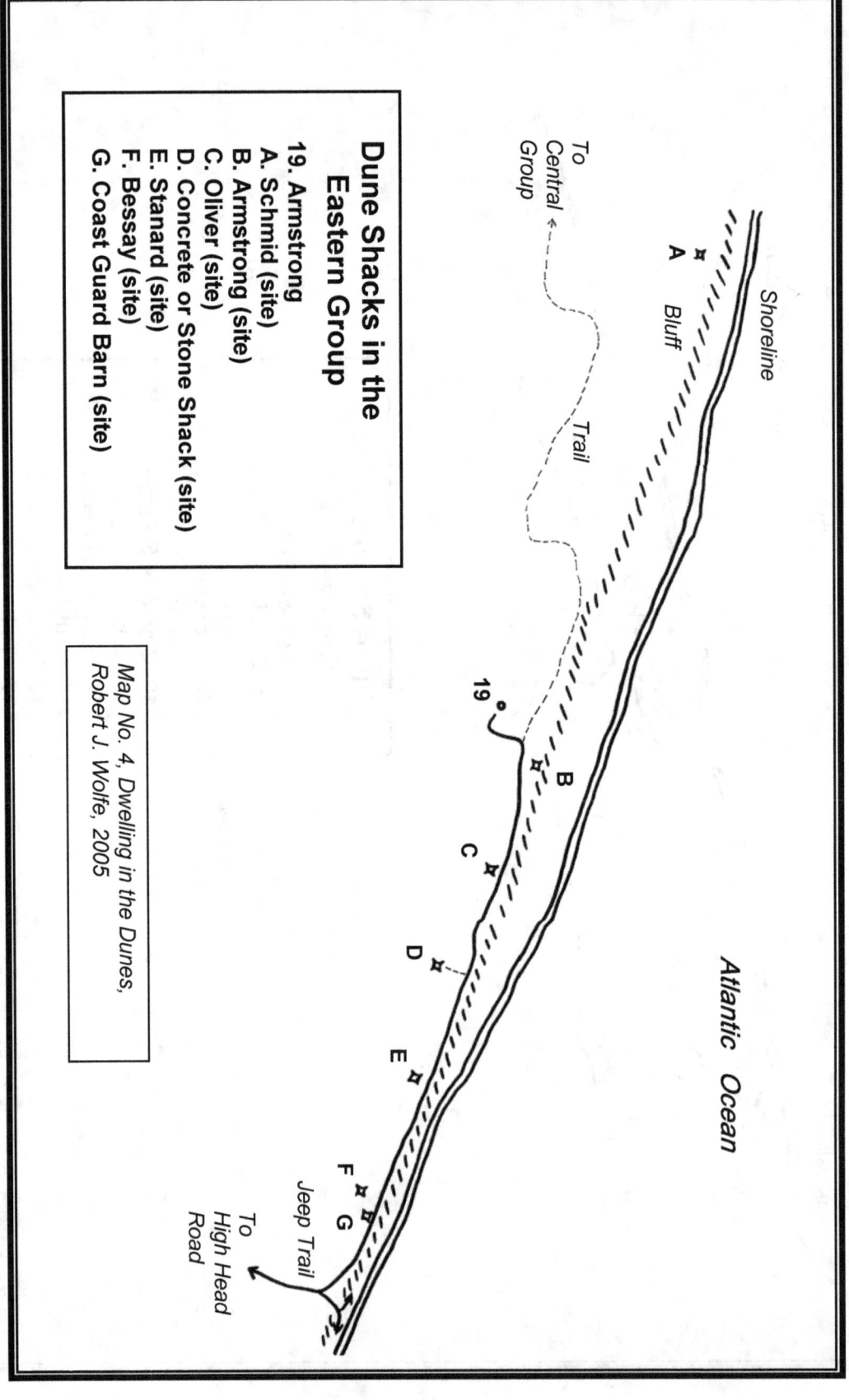

Dune Shacks in the Eastern Group

19. Armstrong
A. Schmid (site)
B. Armstrong (site)
C. Oliver (site)
D. Concrete or Stone Shack (site)
E. Stanard (site)
F. Bessay (site)
G. Coast Guard Barn (site)

Map No. 4, Dwelling in the Dunes,
Robert J. Wolfe, 2005

To
Central ←
Group

Trail

Shoreline

Atlantic Ocean

A
Bluff

19°

B

C

D

E

F
G

Jeep Trail

To
High Head
Road

Jeep trails were unpaved, single-lane roads over the sand. Before the Seashore, more extensive networks of jeep trails crisscrossed the dunes. The Seashore restricted traffic to a few main trails, providing permits only to dune shack residents and a commercial dune taxi service in Provincetown that offered guided excursions for summer tourists. A padlocked chain blocked the jeep trails, labeled "fire roads." Going and coming, dune residents stopped their vehicles, unlocked the chain, drove through, and secured the chain behind them. Formerly, the old networks of jeep trails connected all the dune shacks. Currently the jeep trails did not directly connect the three shack groups. Portions of the old trail system were gone, eroded or buried. At one place, the trail left the dunes to the beach. A portion of the beach route was seasonally closed to protect nesting plovers and terns, reopened later to over-sand vehicles with permits.

In 2004, there were five or six shacks in the western group, depending on how you counted them (Map No. 2). I counted "six," including the "guest cottage" next to the Adams shack. It's not counted as a separate shack in some listings, leading to an occasional confusion as to whether the current number of dune shacks was eighteen or nineteen in total. My total count was nineteen dune shacks in use in 2004. On these maps, I've identified each shack by its current shack heads, unless a building name had supplanted a shack head name. For example, I've designated the Beebe-Simon shack after its two current heads, Emily Beebe and Evelyn Simon. Similarly, the Malicoat-Lord shack is identified by its two heads, Conrad Malicoat and Anne Lord. By contrast, C-Scape (a building name) seems to have supplanted past family names attached to that shack, so I have designated it the C-Scape shack. The Champlin shack (headed by Nathaniel and Mildred Champlin) was sometimes called Mission Bell, so both names appear in my designation. My system here is for clearly counting and identifying shacks. I don't intend my designations to replace customary naming practices of dune dwellers, which are discussed in a later chapter. Whoever might attempt to conform shack names to my system, or to any single standard naming system, will likely incur the wrath of some dune dwellers who care deeply about certain shack names.

In 2004, there were twelve shacks in the central group, located in Map No. 3. The map also identifies the ruins of the Peaked Hill Coast Guard Station, its third historic location, near the center of this cluster of shacks. A major footpath provided access between the central group and town, formerly an extension of Snail Road into the dunes once used by vehicle traffic.

In 2004, there was a single shack (the Armstrong shack) in what was once an eastern group of seven shacks, shown in Map No. 4. This shack was moved away from the spot on the eroding bluff noted as the "Armstrong site." Map No. 4 indicates the approximate locations of five shacks that no longer existed in 2004 (the Schmid shack, Oliver shack, Concrete shack, Stanard shack, and Bessay shack).

My set of maps do not show the locations of four other former shacks. The Vevers-Pfeiffer-Giese shack location in the southern group falls just outside the map boundaries. The Ford shack in the central group burned down and its location near the Clemons-Benson shack is not shown. An old coastguard boathouse used by the Malkin-Ofsevit-Jackson family and others as a shack fell into disuse and was demolished by the Park, its location near the station ruins not shown on the maps. And I was told of another inland concrete shack, a "work of art," whose fate and location are not known to me. If one includes these additional places, I count a total of twenty-eight dune shacks in use at the time the Seashore was created in 1961. However, I would not be surprised if I have missed one or two more.

Sources and Methods

The primary sources of information for this ethnographic report were long-term shack residents. While the dune shacks may be perceived as "old people" that "collect stories," it's the dune dwellers who tell them. So I sought out shack experts to interview. I also walked the dunes, at times guided by experts to interpret what I was seeing. I took pictures. I scribbled notes. And I sketched maps. The stock tools of cultural anthropology are direct listening and observing, ears and eyes being the anthropologist's main instruments, a method called "participant observation." This was my approach. Afterwards, I transcribed hours of recorded conversations with shack residents, forming the primary basis for this report.

I traveled to the lower cape four times to collect information. In mid-July of 2004, I made an introductory visit accompanied by Chuck Smythe, Ethnography Program Manager for the Park's Northeast Region. He introduced me to staff at the Cape Cod National Seashore headquarters in Wellfleet, including the acting superintendent. On this three-day visit, Sue Moynihan, Chief of Interpretation and Cultural Resources Management, and Bill Burke, Cultural Resources Program Manager, gave me a guided tour of the dunes. Driving the jeep trails, they introduced me to the shacks. I returned in early August to spend an entire month of research. I did the bulk of work then, finding experts, conducting interviews, and making observations. At night I slept at a Seashore cottage in Eastham, shared with several other seasonal researchers. After a short break in early September, I returned again in mid September to interview experts available only after Labor Day. From October 2004 through April 2005 I organized materials, transcribed tapes, analyzed information, and developed my findings. The information I examined included documents provided by the Park, collected from Seashore files by Chuck Smythe. In late April I returned to the lower cape for a final research session with dune shack residents interviewed the previous summer. Before a group of about thirty shack residents, I previewed the preliminary findings so that I might identify conspicuous gaps or errors before the conclusion of writing.

Interviews with shack residents were extraordinarily productive. Every shack head I identified for a taped interview was available and willing except for one, whose health precluded a formal interview. In this endeavor, I began with a list of shack heads from the Seashore. The list contained names, addresses, and phone numbers of occupants with legal relationships with the Seashore. To each legal occupant I mailed a letter that introduced the project, requesting interviews during August. Accompanying this was a letter of support from the Seashore. The Seashore also released a public announcement about the project. I began phoning shack heads after I arrived in person on the lower cape to set up interview appointments. At the time of an interview, shack heads commonly volunteered names of additional people that I might contact. A few shack users contacted me directly, asking for interviews. I also compiled names from the written testimonies and letters in Seashore files. In this way, I developed a list of people who might be interviewed.

By the end of research period, I completed formal interviews with 47 shack residents. The people I interviewed are listed in the following table. Every shack was covered by a formal interview except one, the Wells shack. However, I learned a bit about this shack through an informal session with Ray Wells, its shack head. Of those interviewed, there were 25 females and 22 males. Their approximate ages fell in the following age ranges: 20-39 years (6 people), 40-59 years (24 people), and 60+ years (17 people). Regarding residency, 21 persons had a house in a lower cape community, 25 persons had a house off Cape Cod, and one person lived year-round at his shack. Of this group of shack residents, 20 were shack heads, 11 were relations of shack heads, and 16 were friends of shack heads. Finally, of this group, 39 persons were

Shack Residents Formally Interviewed

Count	Date	Shack Resident	Principal Dune Shack(s)
1	9/12/2004	Richard D. Arenstrup	Armstrong
2	9/12/2004	Constance Armstrong	Armstrong
3	9/12/2004	David G. Armstrong	Armstrong
4	9/12/2004	Janet Armstrong	Armstrong
5	9/12/2004	John Cheetham	Armstrong and others
6	8/21/2004	Lawrence Schuster	Schuster
7	8/10/2004	David Adams	Adams
8	8/10/2004	Marcia Adams	Adams
9	8/10/2004	Sally Adams	Adams
10	8/10/2004	Andrea Champlin	Champlin
11	8/10/2004	Mildred Champlin	Champlin
12	8/10/2004	Nathaniel (Nat) Champlin	Champlin
13	8/10/2004	Paul Champlin	Champlin
14	8/10/2004	Maia Champlin Peck	Champlin
15	9/13/2004	Josephine Del Deo	Schnell-Del Deo
16	9/13/2004	Salvatore Del Deo	Schnell-Del Deo
17	8/20/2004	Tom Boland	C-Scape
18	8/20/2004, 8/25/2004	Jay Critchley	C-Scape
19	9/14/2004	Bill Fitts	Werner (Eurphoria), Werner (Thalassa), Gelb-Margo-Zimiles
20	9/14/2004	Harriet (Hatty) Fitts	Werner (Eurphoria), Werner (Thalassa), Gelb-Margo-Zimiles
21	8/27/2004	Joyce Johnson	Various shacks
22	8/17/2004	Genevieve Martin	Various shacks
23	8/6/2004	Julie Schecter	Werner (Eurphoria), Werner (Thalassa), Gelb-Margo-Zimiles
24	8/17/2004	Emily Beebe	Beebe-Simon
25	8/8/2004	Marianne Benson	Clemons-Benson, Fowler
26	8/8/2004	David Andrew Clemons	Clemons-Benson, Fowler
27	8/8, 8/23	Peter Clemons	Clemons-Benson, Fowler
28	8/8/2004	David Forest Thompson	Clemons-Benson, Fowler
29	8/28/2004	Marsha Dunn	Dunn
30	8/28/2004	Scott Dunn	Dunn
31	9/16/2004	Anne Lord	Malicoat-Lord
32	9/16/2004	Conrad Malicoat	Malicoat-Lord
33	8/29/2004	Dawn Zimiles	Gelb-Margo-Zimiles
34	9/17/2004	Martha Rogers Zimiles	Gelb-Margo-Zimiles
35	9/17/2004	Murray Zimiles	Gelb-Margo-Zimiles
36	8/26/2004	Irene Briga	Various shacks
37	8/13/2004	Samuel Jackson	Jackson
38	8/13/2004	Zara Jackson	Jackson
39	8/11/2004	Don Brazil	Tasha
40	8/4, 8/7, 8/9, 8/18	Paul Tasha	Tasha
41	8/11/2004	Paula Tasha	Tasha
42	8/22/2004	Maureen Joseph Hurst	Tasha
43	8/22/2004	Susan Leonard	Tasha
44	8/22/2004	Kathie Joseph Meads	Tasha
45	8/22/2004	Theo Cozzi Poulin	Tasha
46	8/18/2004	Gary Isaacson	Isaacson-Schecter
47	8/18/2004	Lauri Schecter	Isaacson-Schecter

primarily connected to shacks with personal reservations or leases, while 8 persons were primarily connected to shacks with reservations or leases with a non-profit organization. With this set of people, I recorded 2,606 minutes of interviews (43.4 hours). This material was the primary source of information on the cultural patterns of dune shack residents.

Frequently, several shack residents interviewed together, as shown by the dates in the table. The largest group was an interview with eight members of the Adams and Champlin families, who had requested a joint interview. Other interviews were with co-heads of shacks, such as Scott and Marsha Dunn (Dunn shack) or Jay Critchley and Tom Boland (C-Scape shack). Some interviews were with a single person, such as Emily Beebe (Beebe-Simon shack) and Julie Schecter, a leader with Peaked Hills Trust, a nonprofit organization connected to the shacks. Several interviews took place at dune shacks. Other sessions were held at restaurants or homes in Provincetown, Truro, or Wellfleet.

Interviews followed a semi-structured format. I asked general questions, guided by a list of topics I was interested to learn about. Shack residents gave answers while I recorded. Periodically, I probed for detail or clarification. Then I would ask another general topical question and the semi-structured conversation continued. While exploring common subject areas with all shack residents, every interview was slightly different in the order and specifics of questions, tailored to the resident's expertise and interests and the information gaps I needed filled. In many cases, shack residents came into interviews primed to inform me about certain subjects. The sessions were specifically designed to allow for this. With group interviews, the information commonly emerged from lively conversational exchanges between shack residents, participants responding to one another's points. With few exceptions, shack residents appeared eager to inform me about the dune shacks, their histories, and their uses. A substantial volume of information was shared in a short amount of time under these conditions.

During select interviews when there seemed to be sufficient time, I collected information on cultural sites in the dune district. In this semi-structured format, I asked shack residents to look at aerial photographs and U.S.G.S. topographical maps of the dune district. I asked shack residents to locate places of significance to them, along with the places' names and reasons of significance. Sites were marked on transparent plastic sheets placed on top of the aerial photographs. This methodology provided information for compiling maps of cultural sites.

On an informal basis, I spoke with many other people during my research on the lower cape. I acquired considerable basic information from these conversations, such as local history, seasonal patterns of activities, gossip about colorful residents, and so forth. I spoke with art gallery owners along Commercial Street in Provincetown, learning about the art trade. I chatted up a vacationing social worker washing clothes at a Laundromat, learning about summer cottage rentals. I spoke with rangers near Race Point, learning about law enforcement on the beaches. I commiserated with commercial fishermen in Provincetown pubs to learn about fishing, and with street musicians to learn about expatriate students from Eastern Europe providing cheap summer labor in the tourist shops. I developed a sense for the shape of life on the lower cape by living there, participating, and observing for this short period.

Scope and Limitations

At the start of this project, one dune shack resident told me politely, during our first contact by phone, that she hoped the Park Service had hired someone with "the proper frameworks" to conduct the ethnography. I'm not certain she was entirely satisfied after meeting me. I am trained as a cultural anthropologist with twenty years' experience documenting traditional cultures on public lands in Alaska. Given my background, many aspects of local culture on the lower cape were substantially new for me. In particular, the expressive fine arts culture and art colonies were new, requiring quick study. In total, I spent less than two months physically on the lower cape conducting interviews and making observations. Given the constraints of research and my particular strengths and weaknesses, I expect some experts will find certain deficiencies in the report. I may have gotten some things wrong, not hearing correctly or observing completely. For such errors in the report, I must take full responsibility. But in consolation, I remind myself that ethnography is science, amenable to correction with additional information. Scientific description and explanation should always be considered incomplete, needing expansion and refinement, challenges for future work.

The ethnography has one limitation that may be identified upfront, one purposively built into the design of the project. I was not charged with conducting formal interviews with the Park Service. By design, such interviews were outside the scope of work for the project. Accordingly, I did not conduct formal interviews with members of the Cape Cod National Seashore to collect information related to the dune shacks. This constraint exposes the final report to the potential criticism that it's "one-sided" in that it presents information from shack residents but not shack owners. If there are two sides to some stories, one of the sides may be missing. In digesting the findings of the report, readers should keep this limitation in mind. The information about dune shack culture primarily comes from long-term dune shack residents, the focus of the project. It is an ethnographic description of traditions and cultural patterns from people who have lived in shacks.

A final caution is important concerning perceptions of bias. A good ethnography seeks to describe the world of a group of people, not just as the outside social scientist sees it, but also as the subjects themselves portray it. When a cultural anthropologist describes something, it does not mean the researcher necessarily is agreeing with it, or sympathizing with it, or advocating for it. Readers of ethnographies sometimes jump to that erroneous conclusion, that detailed ethnographic description is the same as subjective agreement, distorted by personal bias. In this report, I have tried to paint an accurate picture of the traditions, culture, beliefs, values, and customary practices of dune shack residents. Frequently, I have tried to express them using the words of the dune dwellers themselves, transcribed from recorded interviews. I believe recorded interviews can present information with authenticity, straight from the source. Extensively quoting a dune shack resident is done for rich description. It should not be misinterpreted as advocacy, a soapbox masquerading as science. It's true that this report places great trust in the dune shack residents to present authentic pictures of their lives as they experience them. It's also true that this is done with the goal of presenting rich descriptions of cultures and traditions.

Acknowledgements

There are a great many people who provided much-appreciated help in the project. Literally, this project could not have happened without them. In this short section, I'd like to give my thanks for this assistance. I've never conducted a research project with such a uniformity of support as this one. I can truthfully say that this was my first project where people actually found me out, begging to be interviewed. In social science, it's generally the researcher who begs.

My first thanks are extended to the forty-seven dune shack residents who provided formal interviews. They are listed above, individually. They took me into their homes, fed me, and buried me with detailed information, much of it very personal, most of it exceptionally useful.

My hope is that this group finds that the report does justice to the body of knowledge they shared. Among this group, I'd like to especially thank Paul Tasha and Debra Pelletier, whose cool lemonades at their home on Tasha Hill in Provincetown saved me from heat exhaustion several times. Paul Tasha gave me several valuable guided tours through the dunes, north Truro, and Provincetown, volunteering a lifetime of expertise on these places. A special thanks is extended to Peter Clemons and Marianne Benson for the use of their Backshore Art Gallery as an occasional place of rest between interviews in Provincetown, and for compiling extensive historic materials on the shacks for my use. A special thanks is given to Josephine C. Del Deo for the use of her historic materials, particularly her manuscript, "The Dune Cottages at Peaked Hill Bars," written by her in consultation with Ray Martin Wells, Philip S. Packett, and John Corea, a basic source on shack history, and her book, *Figures in a Landscape, The Life and Times of the American Painter, Ross Moffett, 1888-1971*, containing an account of the preservation of the Province Lands. Others also graciously provided materials for the report. Dawn Zimiles provided digital photographs of artists and writers connected to her family's shack. Murray Zimiles provided materials on the work of Boris Margo and other family members. David Armstrong provided historic materials pertaining to the Great Beach Cottage Owners Association. Jay Critchley and Tom Boland provided statistical information on the users of the C-Scape Shack, as well as logbooks of its occupants. Joyce Johnson provided information on the Truro Center for the Arts at Castle Hill. John Cheetham provided transcripts of interviews he conducted with shack residents.

Throughout the project, I received tremendous help from personnel in the National Park Service. Of particular help were Sue Moynihan, Chief of Interpretation and Cultural Resources Management, and Bill Burke, Cultural Resources Program Manager, the Seashore's resident staff historian. They were available whenever I had a need, providing me with materials, a telephone in the headquarters' library, an introductory guided tour of the dunes, and answers to questions based on their wealth of personal expertise. Mark Adams, GIS Specialist, and Elizabeth Murray, Cartographic Technician, put together aerial photographs and topographic maps of the dune district that were essential for the mapping of cultural sites. The base photo was from the Commonwealth of Massachusetts, MassGIS (2001 digital orthophoto). General and administrative help was provided by Marianne McCaffery, George Fitzgerald, and Nancy Finley. The Cape Cod National Seashore provided me with accommodations at a cottage in Eastham for the duration of the fieldwork. And finally, Chuck Smythe, Ethnography Program Manager, Northeast Region, National Park Service, Boston provided an extremely supportive work environment for the project. He compiled a wealth of documents from government records to give me a solid background to the dune shack issue. The credit for the conceptual and fiscal foundations for the project belongs to him.

Chapter 2. The Roots of Tradition

Dune shack residents today recount a history of the shacks covering over a hundred years, from at least the late 19th century and stretching into the 21st century. The traditions associated with the dune shacks of today are rooted in that history. The dune shacks figure in the history of Provincetown and lower Cape Cod, and have national significance in the history of American literature, theater, and art. The history of the shacks and the evolution of their distinctive uses are marked by significant periods, or milestones, for the shacks and the people who have used them. This chapter traces the history of the dune shacks organized around these periods.

Of course, history is more than just a chronicle of milestones. For living traditions, history also provides sources of present customary patterns. In history is found an understanding of the present, origin stories that make sense of things today. It provides a logic for the evolution of the past into the present, a rationale for why things are the way they are. It offers historic figures and social groups with whom people of the present may identify. Because of this, the past establishes a foundation for the actions of people who choose to emulate and build on it. Histories influence the future. The history of the dune shacks, as told by current shack residents, does all of this.

One powerful origin story celebrated by the townspeople at Provincetown is that of the Pilgrims of the Plymouth Colony. In the national consciousness, the story of this one colony has been elevated to mythic proportions. It's taught to children in public schools, commemorated in a major national holiday, and in Massachusetts, sold to tourists to generate substantial income. Provincetown was the original landfall of the Plymouth colonists in 1620, where the Mayflower Compact was signed. The town has staked a claim in this mythic tale. Provincetown's Pilgrim Monument, America's tallest granite tower, consecrated by Presidents Roosevelt and Taft in grand celebrations, commemorates local events. At a smaller scale, Sunny Tasha and Harry Kemp, noteworthy dune shack residents, organized annual public reenactments during the 1950s of the Pilgrims at Provincetown. Dressed in period costumes, they commemorated the "First Wash Day" by women off the boat. The homespun ceremony reaffirmed the humanity of this mythic tradition. Tasha and Kemp's reenactments, drawing in family members and neighbors, made the Pilgrims' story their personal story too.

Like the Pilgrims for Americans, dune shack history captures the imaginations of contemporary dune shack users. The historic stories are recounted with pride and passion. They are very personal origin tales whose retellings provide grounding for contemporary dune shack society. Dune shack residents I interviewed did not trace their origins to the Plymouth colony. Shack users of today were not modern-day Pilgrims, though the early colonists set the stage. I was told that early colonists on the lower cape inadvertently unleashed the dunes by cutting and overgrazing. But none of the dune shack residents I interviewed started dune shack history with the Pilgrims. Instead, dune shack residents located their origins in Yankee and Portuguese fishermen, in the lonely surfmen stationed on the Backshore, in the occasionally-renowned but typically-starving writers and visual artists from the Provincetown art colony, and in a succession of unconventional mavericks experimenting with life on the dunes. Hatty Fitts, a Provincetown resident whose family has used dune shacks for four generations, provided a thumbnail synopsis of this history during our interview:

> Most of the shacks originally were built by the precursor to the coastguard for their families or for their own use. Then the local fishermen started using the shacks because it was far more convenient staying out there to fish off the Backshore than it was to keep trekking

across. Then it was by word of mouth and that sort of thing that the artists that were coming into town at that time found out about the shacks, especially writers. Through their friendships with fishermen and others, they started using the shacks.

This chapter organizes what the dune dwellers recounted about their history, interweaving their stories to provide a glimpse of the origins of dune shack society. The living traditions on the dunes described in later chapters find their roots in this history.

Shipwrecks and Life Saving Stations

Many dune shack residents like Hatty Fitts traced their origins to the lonely surfmen of the lifesaving stations on the Backshore, the precursors to the coastguards. The surfmen watched the offshore shoals for ships in distress, providing rescue and shelter for people shipwrecked by storms. The U.S. Life Saving Service began in 1872 and was made part of the newly-formed U.S. Coast Guard in 1915. On the lower cape, there were stations at Race Point, Highland, Peaked Hill Bars, and High Head, with lifesaving huts between stations. Lifeguards patrolled the beaches from the stations to the huts and back. The lifesaving huts evolved from the so-called "charity huts" of the mid-19th century, erected along Cape Cod beaches by the Massachusetts Humane Society as shelters for the shipwrecked.

Kathie Joseph Meads, a dune shack user, saw a direct linkage between the lifesavers and the dune shacks. Sunny Tasha, the matriarch of the extended Tasha family and friend of Harry Kemp, the poet, taught her this connection:

> Sunny Tasha always said, "The oldest tradition here in these dunes and these shacks was that of the lifesaving stations." She wanted to run this shack as one of the lifesaving stations were run, which was with an open door, with whatever a stranded sailor at sea, who was in a shipwreck, would need. He'd make his way up to one of these little dune houses and he'd open the door. There'd be a set of matches and a candle. There'd be some sort of little food and shelter from the storm. The coastguardsmen, the lifesaving service, would make their way from one shack to another. These were always open. More than likely they'd be by themselves. But they sometimes found a shipwrecked sailor there and brought them back. That's what the tradition of these shacks was. And it remains. And it should be carried on in that tradition. It didn't start with Sunny. It was what these shacks were meant to be way, way back.

Paul Tasha confirmed that the open-door policy of the Tasha shack was connected with the surfman tradition, as shelters for wanderers in need, through Harry Kemp and his mother:

> It's the old school of thought from Harry Kemp and my mother. They wanted anybody who was wandering in the dunes, if they needed shelter, to be able to come in. So there was never a lock on the door. And to this day, believe it or not, in this day and age, there's still been no locks. We get people from all over, all over the world really.

These accounts illustrate how the oral traditions passed down within this family line have directly influenced use patterns of the Tasha shack today. The shack's open-door policy was rationalized by the Tashas as an extension of the traditions of the Backshore lifesavers, who originally built the Tasha shack. After winning the lease for the former Peg Watson shack, Gary Isaacson and Laurie Schecter also decided to follow an open-door policy for their shack, even though this involved risks. Many I interviewed spoke in admiration of the open-door policy of the Tashas,

Isaacson, and Schecter, although they did not follow it. How the open-door policy fits with the pattern of use of shacks is described more fully in Chapter 5.

Like the Tashas, Russell Powell and Julie Schecter connected dune shack origins to the early charity huts of the Massachusetts Humane Society, described in their film, *Shack Time*, which they produced in 2001. The film quotes Thoreau, from his essay, *Cape Cod*, where he describes a lifesaving hut he observed in 1851. Thoreau approved of the general concept of lifesaving huts presented in the Humane Society's tract, but he was not too impressed by the accommodations in the one hut he examined:

Keeping on, we soon after came to a Charity-house, which we looked into to see how the shipwrecked mariner might fare. Far away in some desolate hollow by the sea-side, just within the bank, stands a lonely building on piles driven into the sand, with a slight nail put through the staple, which a freezing man can bend, with some straw, perchance, on the floor on which he may lie, or which he may burn in the fireplace to keep him alive. Perhaps this hut has never been required to shelter a ship-wrecked man, and the benevolent person who promised to inspect it annually, to see that the straw and matches are here, and that the boards will keep off the wind, has grown remiss and thinks that storms and shipwrecks are over; and this very night a perishing crew may pry open its door with the numbed fingers and leave half their number dead here by morning... These houses, though they were meant for human dwellings, did not look cheerful to me. They appeared but a stage to the grave.

What kind of sailors' home were they? This "Charity-house," as the wrecker called it, this "Humane-house," as some call it... had neither window nor sliding shutter, nor clapboards, nor paint. As we have said, there was a rusty nail put through the staple. However, as we wished to get an idea of a Humane-house, and we hoped that we should never have a better opportunity, we put our eyes, by turns, to a knot-hole in the door... We discovered that there were some stones and some loose wads of wool on the floor, and an empty fireplace at the further end; but it was not supplied with matches, or straw, or hay, that we could see, nor "accommodated with a bench." Indeed, it was the wreck of all cosmical beauty there within... We thus looked through the knot-hole into the Humane-house, into the very bowels of mercy, and for bread we found a stone... However, we were glad to sit outside, under the lee of the Humane-house, to escape the piercing wind. (Thoreau 1961:84-90)

None of the current dune shacks was as rustic as the lifesaving hut described by Thoreau. The Tasha shack, the smallest of the dune shacks, approximates the dimensions of Thoreau's charity-house, but currently it offered many more amenities, including windows, double French doors, a bed, well-stocked book shelves, wood stove, and storage attic (see Fig. 20). Among the current set of dune shacks, the Armstrong shack and Dunn shack were said to be located at the sites of lifesaving huts, and the Schnell-Del Deo shack was built upon the second Peaked Hill Coast Guard Station foundations.

The linkage of the dune shacks with the Backshore surfmen is structural, as well as conceptual. As documented by Josephine Del Deo, a resident of the Schnell-Del Deo shack, many contemporary dune shacks in the central district have descended from buildings constructed by the lifesavers stationed at the Peaked Hill Station (Del Deo 1986). The first uses of the dune shacks in the central district were by the surfmen and coastguards, connected with the activities of the Peaked Hill Station. These links with the lifesavers are still within the memories of the two oldest shack residents, Zara Jackson and Ray Wells.

Certain coastguards working at the Peaked Hill Station (Captain Frank L. Mayo, Frank Cadose, Raymond Brown, Joe Madeiros, Louis H. Silva, Louis Spucky, and perhaps a couple others) built several small cottages near the station to house family and other guests, and to rent for income. According to Josephine Del Deo (1986), the cottages were built from loneliness, as places to put loved ones visiting the Backshore from Provincetown:

> The several buildings in this complex were variously service buildings for the station and/or small cottages constructed by the Coast Guardsmen themselves for use by their families in the summer months, loneliness being a compelling stimulant to this activity.

This has become part of local oral history, as retold by Peter Clemons, a dune shack resident:

> There were a couple of cottages built in conjunction with the lifesaving station. They were specifically built for families to come out and visit. This is part rumor, part truth, I suppose. They were built for kids and mothers to come and be somewhere in the vicinity of the guys that were running the coast guard station.

According to this oral history, the shacks were not built for solitude, but to relieve it. Relatives and friends from Provincetown who walked out Snail Road to visit the coastguards at Peaked Hill Station or to fish the Backshore could stay in the shacks. A variant of this pattern of use continues today. Relatives and friends come out to the Backshore to visit the core residents of the dune shacks. This continues to be a primary use pattern of dune shacks, as described in subsequent chapters.

The owners of the small cottages also rented to interested parties, including Harry Kemp, Ray Wells, Hazel Hawthorne Werner, among others. Very early on, several cottages changed owners for modest sums. And the small cottages were moved as storms undermined the beaches. The changing configuration of structures around the Peaked Hill Station was documented by Hazel Hawthorne Werner, an early shack renter and shack owner in the central district. About 1960, Werner drew a series of five maps of the district for Al Fearing, another shack owner (Werner 1971). Based on memory, her maps depict structures at successive time periods, labeled 1920, 1933, 1939-1946, 1946-1950, and 1950-1960. The maps graphically depict how buildings constructed by coastguards at the Peaked Hill Station evolved into a clustering of dune shacks. A summary of the historic periods is as follows, extracted from the Werner maps, supplemented with information from Del Deo (1986).

Werner's 1920 map identifies nine structures at that time:
- The "new" coastguard station (eventually relocated further inland, then decommissioned, and then burned);
- The "old" coastguard station used as the Eugene O'Neill summer home (eventually lost to the sea);
- Captain Frank Mayo's cottage (eventually lost to the sea);
- Frank Cadose's cottage (originally the station's hen house, eventually claimed by Frank Henderson and rented to Ray Wells, then rented to Harry Kemp, becoming the Kemp shack and then the Tasha shack);
- Chief of Police Charles Rogers' cottage (rented to Harry Kemp, eventually becoming the Malkin-Ofsevit-Jackson shack);
- The coastguard boathouse (eventually becoming the second shack of Malkin-Ofsevit-Jackson, and then incorporated as salvage into the Isaacson-Schecter shack); and
- Three sheds near the "old" station.

Werner's 1933 hand-drawn map identifies four additional structures:
Raymond Brown's cottage (eventually becoming the Fearing-Fuller-Bessay-Clemons-Benson shack);
Joe Madeiros' cottage (eventually abandoned and collapsed);
Louis Silva's cottage (eventually becoming the Werner shack named Thalassa); and
A shed built for O'Neill (circa 1921-26)

Werner's 1939-46 map identifies six additional structures:
Frenchie Chanel's cottage (started in 1946, eventually becoming the Schnell-Del Deo shack);
Boris Margo's cottage (rebuilt several times, eventually becoming the Gelb-Margo-Zimiles shack);
Hill-Ford cottage (eventually accidently burned);
Stan Fowler cottage (the Fowler shack);
Hazel Hawthorne Werner cottage (originally built by a coastguard Louis Spucky for his wife and bought by Werner in 1946, eventually becoming the Werner shack named Euphoria); and
A coastguard watchtower (removed from the station and relocated to the beach, eventually destroyed by fire in 1946).

Werner's maps of the next two periods (1946-1950; 1950-1960) chronicle the movements of the station and shacks in the central district, but mention no additional shacks being built (except the Gelb-Margo rebuilds). However, in addition to the above shacks illustrated in the Werner maps, Del Deo (1986) reports that coastguards P.C. Cook and Joe Medeiros built another cottage just east of the station in 1931 and sold it to the Braatens soon after, eventually becoming the Schuster shack. Conrad Malicoat reported that the Schmid shack originally was built by a coastguard named Meads, and owned by Peg Watson. Other shacks outside the central group also may have built by coastguards.

For about two decades (circa 1919 to 1939), dune shack residents and the coastguards formed an interactive community in the central cluster of shacks. They lived as neighbors, providing favors for one another. This Backshore community was described by Zara Jackson, who remembered it as a young child about 1929:

The Hill [Peaked Hill] was basically where Margo's [shack] is. It was a different configuration then. There was the top shack, the middle, and then Brownies' [Raymond Brown's shack]… There was Louis Spucky's shack that Hazel owned, and then a middle one, and then what became the Fearing shack. The Hill was immediately parallel to the station. Remember, all of these shacks, with the exception of mine, were built by the coastguards for their families. They would sometimes bring their families out, or they'd have a little place to go themselves. So that was like a little community, which was a different thing from when people later wanted to come out only to be alone.

When the station was opened, in the early years, these men all came from this region. They were from Provincetown, or very close to it. They would walk home on liberties. For a weekend or whatever time they would get, a day-and-a-half, or whatever, they would walk back to town. Now there's a little bit left of the old Snail Road which came over the dunes. It was completely lined with planks for walking. They would walk back carrying their liberty bags. The Snail Road through the low-lying woods could accommodate a horse and wagon. "Greeny" Silva, I think his name was Emmanuel, was the cook. He'd go in weekly for supplies. Sometimes we'd get a ride with him. Otherwise we'd walk into town and carry things back in an onion sack. We'd pump cold water and keep it for a few days in a bucket of cold water. That was the early preservation.

The coast guard station started out with a horse and wagon. They had scythes and they cut the beach grass for Betsy's hay for the winter. Then they got a tractor. And then they had a truck. The truck was in the middle 1930s. Some people had flivvers. Jimmy Thomas had a flivver, an old car like a Model-T that had been stripped down. They would under-inflate the tires. He would bring it out here. He could make it go almost anyplace. And Frankie Henderson had a car that he got out here.

This was a very old-fashioned culture. In a sense everyone looked up to the coastguards. They were our guardians. Occasionally we'd get invited to have a meal with them. It was a big dining room. They'd have iron stone mugs and heavy dishes from the Navy. There was a big coal stove in the kitchen, and it was always going. There was always a pot of coffee on. This was a different world from today.

When the station was on the bluff, the watch was in the cupola because that overlooked the beach. After the station was moved back to this area, they built a little watchtower, a little watch house, and they'd walk there. Joe Morris was the chief mate at the coastguard station. My mother paid him to watch out for the shack in the winters. Joe Morris would open and close it… While the coastguard station was open, there was no vandalism.

I think the station closed up around 1938 or 1939. Then there was just one coastguardsman, Frank Henderson. He was assigned to watch out for the station and to keep things in order. Then it was closed completely. It was re-opened after World War II, which was 1941. The Coast Guard became a part of the Navy. So it was a very different culture then. The Navy culture was not the original one.

In this oral history, Zara Jackson recalled a type of friendly interaction with the coastguards. She said they were admired, "our guardians." People regularly came and went between the Backshore station and Provincetown. She got rides to and from town in the station's horse-drawn wagon. She occasionally dined with the crew in the station's dining hall. Her mother hired Joe Morris, a station man, to open and close their shack and to watch it over the winter. She recalled that Morris built their shack's outhouse and children's swing set, on hire by her mother. According to Jackson, the dune shack residents and the coast guard station personnel formed a type of community on the Backshore. The involvement of the lifesavers in this community ended with the closing of the stations just before World War II. The station was briefly reopened by the Navy during the war.

The coastguards assisted dune shack residents during storms that threatened the shacks. Zara Jackson recalled the series of storms that undercut her shack and the coast guard station, and that destroyed the old station used by Eugene O'Neill:

When we first came out [1926], our shack was not in its present location. It was all the way in front of where the Boris Margo's shack is, on the bluff. It stayed there until about 1932 or so. When we were first out, it was also the time when there was the old station that Eugene O'Neill had lived in. You know "Reds" – that was a film about O'Neill living out here – that was that period. Well, O'Neill had been here earlier than us. He was here around the period of World War I. By the early 1930s, he had left Peaked Hill. That station was being washed out to sea just about that time. There were tremendous storms in the early 1930s that were eroding the cliffs all along there. So the shack was really in danger.

I have a memory of this one night when Harry Kemp and Hazel Hawthorne and I don't remember who else was living here then. It was sort of like a party in the shack, because there was this enormous storm going on. When we got up in the morning, to get out of the screen door you had to back along the front because within two feet was the drop of the cliff.

The shack was moved back on rollers. They had huge telephone poles. The coastguard pulled it back with the tractor a fair amount. But that was as far as it could go. It couldn't stay there because this erosion was still going on. So it was taken apart into sections by Jesse Meade, the same mover for the station. The station was in jeopardy too. So at just about the same time, they were moved back. They were in a parallel position [with her shack].

So the coastguards helped save the Jackson shack as well as their own station. The old coast guard station used by O'Neill as a summer residence was not so fortunate. It fell into the sea and was lost in 1932.

O'Neill's Summer Residence Falling into the Sea

For some dune residents, such as the Tashas, Jackson, Wells, and Schuster, their shacks provide direct reminders of the lifesavers on the Backshore. By and large, these are positive associations. In local oral traditions, the lifesavers are remembered as brave, dedicated public servants, and good neighbors. However, not all associations are positive. According to Julie Schecter, the connection of Thalassa with the lifesaving period keeps some people from using that shack, the problem being ghosts:

Hazel Werner bought Thalassa from one of the coastguardsmen... There's a story of a shipwreck that took place right out there. They say they laid the dead men out on the floor of Thalassa. There are people who won't stay out there because they can hear the dead men talking to them.

Werner's shack continues to be infused with the spirits from that tragic event, by this account.

Fishing Shacks

Some dune residents trace the origins of dune shacks to fishing by Provincetown residents, to the sheds, shacks, and huts used by fishers for shelter and gear that predate the lifesavers. In her interview, Kathie Joseph Meads traced shacks to fishing huts. Meads said that Thoreau observed fishing huts in his travels around Provincetown in 1851 and 1857:

The shacks did exist at the time of Thoreau, which was before the coastguard. They had fisher shacks out here in the 1800s.... I hope you've gotten a chance to see Henry David Thoreau's *Cape Cod*. There's a chapter at the very end, one of his Provincetown chapters, where he refers to the fisher shacks, which are these dune shacks, where he spent the night... There's a little few sentences where he spent the night in the dune shack, in the fisher shack as he called them. He said everyone should engage in this kind of experience with nature, because this was what was meant by a "true hotel." This is where the environment was brought to you. That's what a true hotel should be.

Thoreau's reference to a "true hotel" is found in the final paragraph of his essays on Cape Cod, just before the famous statement about a man putting all of America behind him, standing on the Backshore and facing the wide Atlantic:

A storm in the fall or winter is the time to visit it; a light-house or a fisherman's hut the true hotel. A man may stand there and put all America behind him. (Thoreau 1961: 319)

So for Thoreau, the "true hotels" for a Cape Cod vagabond like himself were places like the lighthouse on High Head, overlooking the Backshore, and a fisherman's hut on the beach, especially during stormy fall or winter.

Thoreau's essays do not explicitly describe him ever lodging in a fisherman's hut, although he might have. In his account, he refers to "the make-shifts of fishermen ashore," referring to the rough buildings erected in town and near places where people fished. Thoreau describes Provincetown as a collection of houses, fishermen's huts, fish houses, storehouses, and salt works:

The town is compactly built in the narrow space, from ten to fifty rods deep, between the harbor and the sand-hills, and contained at that time about twenty-six hundred inhabitants. The houses, in which a more modern and pretending style has at length prevailed over the fisherman's hut, stand on the inner or plank side of the street, and the fish and store houses, with the picturesque-looking windmills of the Salt-works, on the water side... This was the most completely maritime town that we were ever in. It was merely a good harbor, surrounded by land dry, if not firm, an inhabited beach, whereon fishermen cured and stored their fish, without any back country. When ashore the inhabitants still walk on planks.

As described by Conrad Malicoat, a dune resident (see below), starting in the late 1800s, many poor artists and writers happily adapted the "rinky-dink" shacks in town into seasonal dwellings and studios, bases in their quests as artists in the emerging Provincetown artist colony.

Most of the shacks used today were not originally fishing shacks. However, some may trace their origins to that activity, at least in part. Zara Jackson called Frank Mayo's shack a "little fishing shack," in her description of its demise after a major storm in the early 1930s:

At that time Frank Mayo had a little fishing shack. It ended up cantilevered, half on the cliff and half over the ocean. So the coastguards put a huge rope around it and attached it to their tractor. They were trying to pull it, maneuver the shack back. Instead it sort of just folded up like a house of cards, and went over the brink.

Frank Mayo's shack was one of the shacks in the central cluster, in the vicinity of the Peaked Hill coastguard station. So fishers used one of those shacks, and perhaps others in the central cluster.

As described above, the coastguards employed a different approach using rollers to help pull the Jackson shack away from the eroding cliff edge.

Some Provincetown residents erected shacks to use as fishing bases, a type of fishing camp on the periphery of town. They were not exclusively fishing bases: local residents constructed them also for merrymaking, dancing, and drinking. According to Nathaniel and Mildred Champlin, their shack was built by Dominic Avila, a member of the Provincetown Portuguese community, for fishing and partying in the early 1930s. Champlin himself discovered the Avila shack in 1948, also while fishing and partying on the Backshore with buddies. Other shacks in the northern cluster built by Jake Loring (the Adams shack) and Albert Nunes and Jake Waring (the Beebe-Simon shack) apparently were built for similar purposes. Fishing (surf casting from the beach) was one of the uses of the Backshore at that time, and shacks in the northern cluster were linked to this pattern of use.

Some local place names given by shack residents refer to camping sites of fishermen who erected shelters. For instance, the Champlins identified The Bowls, Second Rip, and Bill's Camp as camping places for fishermen (see Chapter 6, Map 6, Cultural Sites Nos. 11, 13, and 19). Bill's Camp was described by the Champlins as a site between Phil Malicoat's shack and the Peaked Hill area where fishermen erected small temporary shelters and tents as bases for fishing. There may be other sites like this on the Backshore that I did not document in the mapping sessions.

The first shack built by Phil Malicoat in the northern cluster was burned down by fishermen who were using it in Phil's absence, according to Conrad Malicoat's history of his shack:

The original shack my father built. It was very small, something like 12 by 16 feet. He built it in the late 40s, about 1948 or 1949. The way he determined its location was this. He owned a piece of property that at that time was 75 feet wide and went from Bradford Street to the ocean. He determined where he was going to put this shack by looking at a map. It wasn't a topographical map, it was just a map that the government put out. He knew the angle because the deed had a description of the compass needle, so he plotted that across the dunes. The map was clear enough, so he could sort of figure out where it was. The map wasn't topographical or else he probably would have hit it right on the nail.

Later in the 1950s he went to Europe. He let some fishermen friends of his use it. I guess they had a party or something and the place burned down. That happened when he was away. And that was a very good thing to have had happen. When they came back, he and I surveyed it. We got a real surveyor's instrument and surveyed it right directly across. We found that this former building was off by about 600 to 700 feet. So we rebuilt that place according to the new survey.

My dad would share the shack with his artist friends in town here. And he shared it with the fisherman. Well, I guess he didn't share it with them after it burned. *[Laughter.]* Actually, I think he was understanding. He was very understanding. He never complained about it. He had to go fishing himself just to earn a living at one point.

It's possible some of the shacks in the eastern group of shacks also might be traceable to fishing shacks. A shack co-owned by Graham Giese and friends was used for fishing before being undercut by tides and removed by the Park Service. It may be more difficult to reconstruct the history of this grouping, as all except the Armstrong shack are gone.

As shown by this information, fishing from temporary shelters was a pattern on the Backshore since the mid 19th century, and perhaps earlier. Some dune shacks were used as bases for fishing, and so might be considered "fishing shacks." Though connected to shacks, fishing from the Backshore evolved into a use pattern distinct from that of dune shack society. Surf fishers began using converted bread trucks for shelter in addition to tents and temporary shacks, according to the Champlins and Adams. Today, the surf casters find shelter in recreational vehicles camped on the beach under the authorization of sand permits provided by the Park (see Fig. 2). This large social group directly descends from the surf fishing patterns of the early 20th century. The surf casters who use sand vehicles are represented in part by the Massachusetts Beach Buggy Association, many of whom fish. The surf casters interact with the dune shack society, but are distinct from them.

Old Provincetown and Rural Economies

The Peaked Hill Bars Coast Guard Station was connected socially and economically to Provincetown during the late 19th and early 20th century. Most surfmen were Provincetown residents from old Provincetown families, primarily members of the town's Portuguese community. Surfmen and others traveled back and forth between the Backshore station and town for supplies. Shacks were built around the station to accommodate visitors from Provincetown. On liberty, the coastguards walked to town to visit family and friends.

In addition to employment as lifesavers, Provincetown residents used the Backshore and dunes for a number of economic purposes during the 19th century. The beach was regularly searched for salvage from shipwrecks, an occupation described by Thoreau in the Cape Cod essays. Salvaging the Backshore was profitable. Wreckage from ships became the property of finders unless the ship owners employed the salvagers. This traditional use of the Backshore by Provincetown and dune shack residents has continued to the present. Shacks commonly incorporate salvage in their construction, as described in later chapters.

Since its earliest days, the dune district provided certain wild food products for Provincetown families, especially deer, hares, migratory waterfowl, cranberries, blueberries, beach plums, rose hips, and mushrooms. East Harbor, the water body surrounded by dunes to the southeast of Provincetown, provided eels, turtles, and other fish. These family subsistence activities were extensions of the fishing economy of Provincetown. Families ate wild foods gleaned from the sea and the land. The plums and berries were preserved as jams. Fish was smoked. The wild food products were eaten in the home and sold for small amounts of income. Like salvaging, the traditional uses of the dunes for food have continued to the present. The food gathering activities have come to be perceived as emblems of a traditional heritage at Provincetown, particularly of the Portuguese community. The uses are viewed as continuations of valued Provincetown traditions, providing satisfaction to families who identify with that heritage.

The dune district was used for small-scale cranberry farms during the 19th and early 20th century. A portion of the dune district was divided into narrow, privately-held tracts running from the shore of Provincetown Bay to the shore of the Atlantic, for use in cranberry farming and other economic activities. The "spaghetti strips" and their uses were described by Conrad Malicoat and Anne Lord, whose family owned two of the narrow parcels during the mid-20th century:

Malicoat: Originally a lot of people farmed out there on the dunes. One of the biggest farming that they did was cranberries. Pretty close to where we live there's a trail that's called the Whistle Path, because the train came across it. That particular path went from Bradford Street [in Provincetown] all the way over to the Backshore. The farmers in horse-and-buggies would go over the dunes and get to their plot of cranberries. This was quite a flourishing business for everybody. There are a lot of cranberry bogs around here. Those little "spaghetti strips" they call them, they were pretty narrow, so each individual owner had their own little private holdings, for whatever they were doing out there. They were maybe harvesting other things, too. I don't know exactly. Wood was harvested. And you could be fined. I remember this historical fact. If you started harvesting your berries before a certain time that some elder determined every year because of the frost, you'd be fined a dollar.

Lord: Or you'd lose some kind of right. I think some of the cranberry bogs were in the Province Lands and really belonged to the town, before they signed it over to the state. This is going back a really long time ago.

Malicoat: Yeah, we're going back to the 1700s, or the 1600s.

Conrad Malicoat described Provincetown in the late 19th century, when dune shack history emerged, as a culturally-mixed community of old Yankees and Portuguese fishermen, who became friendly to artists and writers:

I'm going to get into a little bit of history, because I think it's important to understand. The working class people who were here originally had a dependence on the sea. The fishermen were a group unto themselves. The Province Lands were the farthest out into the water, and therefore a very valuable place to be in terms of being able to harvest fish. These were Yankees. We could call them Yankees, but probably they were from different countries, yet mostly English in origin.

Then came the Portuguese. The Yankee fisherman had started to disperse. The captains had a harder time finding crews. So they went to the Azores and to the Verde islands, mostly the Azores, and they had them as crew. The crews weren't allowed ashore. But they'd jump ship. Pretty soon this town became a Portuguese community, basically. The politics were run for the most part by Portuguese for some time.

The fishing community of Provincetown was described by Thoreau in the middle 19th century. Fishing fleets and transporters came and went from the deep-water port, carrying herring and cod harvested from the near-shore banks. Salt works lined the Provincetown shore boiling sea water into salt for curing fish, burning wood collected from nearby woodlands. Families moved fish to and from storage sheds each day, laying them out in the sun to cure. Traffic was primarily by foot and boat. Packets ferried fish and supplies from Provincetown to the mainland. The mixture of Portuguese and Yankees in Provincetown during the early 20th century was described by Mary Heaton Vorse's *Time in Our Town*, and also *In the Narrow Streets* by Wes Moffit. Several dune shack residents identified the Vorse book as an accurate portrayal of the Provincetown society they knew, before its changes from tourism toward the end of the 20th century. The town's population was relatively poor, subject to the insecurities of fishing, but hardworking and proud. It was a small, maritime-based community with a rural economy, producing food for home use and sale, making a living from nature. Some of the uses in the dunes today are viewed as a continuation of these small-town traditions, as described in subsequent chapters.

Provincetown also was a community with a reputation for tolerance for the maverick and marginal. Salvatore Del Deo, a dune shack resident, believed the freedom derived from Provincetown's geographic isolation and political autonomy:

> In general terms, Provincetown was a free port. That may have set the tone of this place, before the Revolutionary War. They were isolated from the rest of the nation. There was no conscription here in Provincetown for the military service, because it was not considered really part of the United States. It was that kind of rebelliousness. There was an infamous settlement that developed at Hatches Harbor called Hell Town. That's exactly what it was. It was like a miniature Shang-hai and 'Frisco in the gold rush days, all put together.

Provincetown residents in general, and dune dwellers in particular, identify with the tradition of personal freedom at Provincetown. The town's free spirit predisposed it for acceptance of the free-thinking writers and artists who arrived during the late 19th century.

The Provincetown Art Colony

Toward the end of the 19th century, Provincetown attracted another community of residents in addition to the old Yankee and Portuguese fishers. These were visual artists, writers, and actors from the mainland and Europe. Many were seasonal visitors coming for the summer to work. But others tarried longer, becoming year-round residents. Several schools of art were established in Provincetown, attracting students for summer art classes. In a short time, an art colony developed at Provincetown resembling colonies in Europe. Some dune shack residents considered Provincetown to be America's oldest art colony, extremely influential in the history of American art, theater, and literature. They saw their own activities on the dunes as a continuation of that storied tradition. Josephine Del Deo, a dune resident, summarized this history:

> When the artists came here in the late 19th century, the first was Waterman – he came here in the 1880s. He had come from Africa. He had been doing paintings over there of the dunes. When he came here and saw the dunes, he said, "Oh my God, this is incredible." He went out there and did all these dune landscapes. And he put Arabs in them. Then we had Charles Hawthorne who came in 1899. Since Charles Hawthorne's time, this place has been a major art colony. So naturally the fallout from that would be artists out there [on the dunes]. And writers too. Writers have been big out there. Eugene O'Neill. Edmund Wilson. Hazel Hawthorne. Norman Mailer. Myself. You name it. The painters and writers have always gone there, for the things that the painter and writer essentially value the most, which are privacy, solitude, and the contact with nature.

Artists and writers began using the dunes by the early 20th century, an extension of the colony out onto the Backshore. Eugene O'Neill, the well-known playwright, may have been the first writer, working from the old coast guard station at Peaked Hill during the First World War. But others began using shacks surrounding the station and elsewhere. The dunes provided places of solitude for creative work, and places for creative thinkers to gather for collaboration and revelry. These types of uses of the dune shacks have continued to the present. The history of the development of the art colony at Provincetown, and its connection to the dune shacks and the Yankee and Portuguese communities, was recounted by Conrad Malicoat, a resident artist in Provincetown, whose family was part of that history:

> In the late 1800s, and there were probably some artists here before that, I'm sure, there was one particular man, Charles Hawthorne, I'm sure you've heard of him. He proselytized and

he sold the area. This became a counterpart of a European artist colony. He was so excited about his place because of the light. This was a period of Impressionism, so light particularly was an important thing. He went around the country and got students to come to Provincetown. He had a flourishing class for many, many years. That's what originally drew both my mother and her family, her parents, here, in the very early part of the twentieth century. And then later, in the 1920s my dad came here to study with Hawthorne. So, Hawthorne was a very important influence.

Now, when these artists started coming to Provincetown, they loved it, and some stayed. Where did they stay? You might say they stayed in shacks. And what were the shacks? The shacks were old buildings. You see, the fisherman would not live on the waterfront. They had wharfs, they had many, many wharfs in this town. If you see the old pictures, there are some beautiful pictures of the early fishing fleet. Nobody lived on the waterfront. As I heard them say, when you got off the boat you turned your back to the sea. The fishermen lived on the other side of Commercial Street, not too far back.

On the waterfront there were all these rinky-dink kind of buildings, put up for their value as a place to put fishing equipment and so forth. They were constructed out of whatever they had available. The salt works were on the waterfront side, too. Also, when the artists came here there was an economic demise because of the Portland gale. A lot of the salt works were lost, and a lot of the working shacks were demolished because of the tremendous storm in 1898. A whole shift started to happen. Oil was coming in, so they were not sailing, they were using engines to run around and go fishing. At this particular time, artists discovered this town. When they discovered it, a lot of them decided to stay. They found the dregs of buildings to live in, because they're all poor.

One of the reasons for this is their whole quest. An artist's quest is a spiritual quest. It's not money. Money doesn't come into it. You see a shift in that today, where people are being taught to market their art, and that's a whole different thing. The passion to be an artist was something else. It was a very deep commitment to life, to experimentation, to develop a means of sharing whatever was inside of you, seen through your painting. It was a very dedicated profession. People took it very, very seriously. And believe me, I was a son, with my sister, of a family that felt very strongly that way. And many other men and women that I knew of felt that way.

And so this began to infiltrate out there to the Backshore. This was going on now right up to the 1950s and certainly right through the 1950s. You'd have these people come into Provincetown finding inexpensive places to live, and then if they couldn't even find an inexpensive place to live in town, they'd go out to there and build something and live out in the dunes. Nobody cared. The people in Provincetown looked upon them as bohemians. They called them bohemians. They weren't beatniks or whatever, they were bohemians. That was a description of these people that they couldn't figure out, and they kind of laughed at them, or many of them did. They were sort of a particular class of people all by themselves. They would go out there to the Backshore. That place had a reputation for the… I don't know for what. But these people were taking themselves seriously, too. They were not just a crazy bunch of people. They were people who had very particular objectives in mind when they were out there, on the Backshore.

Conrad Malicoat described the attraction of a place like Provincetown to the individual artist. He stated it was a "colony" in the sense of a collectivity of individuals driven by a passion to express:

It's individuals. You have these individuals who are looked at askance for the most part. You know the history of families getting upset when all of a sudden one of their offspring wants to be an artist – they reject them, and then they're out there. So you have these individuals that have a dream, they have an inspiration, an intuition about their life. So they go forth, but they're struggling. Nobody seems to really want them. They have to develop a talent before they get respected, and that's an arduous thing to do. So [in a colony] you have these individuals together, but every individual doesn't agree with the other one. What unites them is the fact that they're all in the same boat of being sort of disrespected, just a fringe group, really trying to bring out something from within themselves.

Provincetown became a cultural center, a small community that drew people engaged in art, theater, and literature. The artists and writers came to Provincetown for education, inspiration, and work. The creative population came and went from the center, a flow that continues today. The historic importance of Provincetown as a cultural center was described by Anne Lord:

Provincetown became a colony with Hawthorne coming, a very charismatic teacher, bringing people in from around the country, following him, and other teachers who were teaching at the same time, and later, drawing all these people in. Then the First World War came and artists who had been going to Europe couldn't really go any more. A lot of Europeans came here. They shared experiences, taught each other, learned from each other, and worked together. It's been called the Cradle of American Art. It drew people from New York, yes. From Boston, yes. From Gloucester to Rockport. From New Hampshire. From the South. From the West. If you go to the National Museum of Art in Washington, go into any gallery and you'll see: the majority of the artists will have spent some time, some of them years, some of them short periods of time, in Provincetown, Truro, and the environs, all drawn here, but especially to Provincetown.

According to several I interviewed, the Portuguese fishermen were instrumental in sustaining the poor artists and writers who came to the Provincetown art colony. Artists found cheap rents in the buildings owned by fishermen along the shore in town or owned by coastguards on the dunes. They received a bemused tolerance from fishermen about the so-called bohemian lifestyle. And they found inexpensive, at times free, food regularly brought to the docks by the Portuguese fishermen. The supportive relationships were described by Susan Leonard, Maureen Joseph Hurst, and Kathie Joseph Meads, three users of the dune shacks with Portuguese roots in Provincetown:

Leonard: These two groups of people, these two cultures, merged here, the artists and the Portuguese. Really, the Portuguese community sustained those people. They would have starved to death. They would have absolutely starved to death if it hadn't been for the fishermen. It was really considered poor form to say 'no' if someone went to the pier when the boats came in with their catch, if someone asked for fish. You were obligated, morally obligated.
Hurst: You never say 'no.'
Meads: The Portuguese felt that if they said 'no,' they were saying no to Christ, and saying no to charity, and therefore their next fishing trip they would not do well. That was just one of their traditional values.
Leonard: It's because of St. Peter being a fisherman, and St. Peter is the patron saint of all fisherman. That runs really deep.

The art colony came to be an integral part of Provincetown. Its members rubbed shoulders with members of the older Cape Cod groups, the Yankee and Portuguese residents. A sense of pride within all social segments led to a type of social egalitarianism in town, and least in the civic self-perception. The ideal was that rich and poor, creative artist and fisherman, seasonal resident and year-rounder, interacted in daily affairs in the small town. Kathie Meads said it was a town in which men like Walter P. Chrysler, Jr. of Chrysler Motors, who owned the Chrysler Art Museum in Provincetown, exchanged vegetables over the fence with her father, a Portuguese fisherman earning $7,000 a year. It was a town of mixed cultural traditions, face-to-face familiarities, and tolerance.

The Emergence of Dune Shack Society

By the 1920s, a dune shack society distinct from Provincetown was emerging on the dunes, tracing its roots to this confluence of local traditions – surfmen at stations, fishers (and revelers) at cottages, writers and artists in hand-me-down stations or shacks, and salvagers, hunters, and cranberry pickers from town. All were closely connected to Provincetown, socially, culturally, and economically. From the 1920s to the 1960s, a period of American history linking horse-drawn wagons with walks-on-the-moon, the dune shacks housed about three generations of residents and their coteries. With its clusters of shacks, the dunes came to house a series of residents. Distinctive personalities connected with the dunes came to be celebrated in the dune's oral traditions, notables such as Eugene O'Neill, Edmund Wilson, and Norman Mailer, as well as others known locally, like Captain Frank L. Mayo, Frank Cadose, and Joe Medeiros. Several extended family lines became established in the shacks during this period, including the Adams, Armstrongs, Champlins, Gelb-Margo-Zimiles, Malicoats, Malkin-Jackson, and Tashas.

During this time, dune shack society was small-town and personal. People knew one another, often on a very personal basis. Dune dwellers interacted and gossiped about neighbors. Stories circulated among dune dwellers about dune dwellers, especially memorable characters with serious life endeavors, such as Charlie Schmid (Dune Charlie), Harry Kemp (the so-called "Poet of the Dunes," the "Last Bohemian"), Jeanne Chanel (Frenchie), Rose Savage Tasha (Sunny), Margaret Watson (Peg), among others. These stories comprised a corpus for an emerging oral tradition. Long-lived dune residents like Hazel Hawthorne Werner, Ray Wells, and Zara Jackson testified to these people and events. In my interviews, dune shack residents frequently gave personal thumbnail sketches of dune notables as iconic illustrations of the distinctive characters connected with the dunes. Talking about the dune shacks meant talking about the people in the dune shacks. Personal sketches provided by Josephine and Salvatore Del Deo illustrate how dune residents commonly talk about dune shack society. At this point in their interview, the Del Deos were discussing dune dwellers other than just artists:

Charlie Schmid
Josephine: He's a wonderful example. Charlie Schmid came here because his wife died. He was from the south. He was so devastated by her death that, just short of dying himself, he wanted to reject the world, forget about it. He came here and built this shack. There was a shack already there, but he built on it, and under it. It defied gravity. He lived there as a complete character until he died, wanting to be alone and left alone to be sad and suffer and whatever else he did. And he drank a great deal, of course. But his contribution was, he did this marvelous study of the swallows. He was so original with it, and so persevering, they asked him to come to Switzerland to a major ornithology conference. He's an example of someone out there who had nothing to do with the arts. But he was driven to being there. And so grateful all his life to being out there. He just worshipped the ground he walked on.

Peg Watson

Josephine: When Peg Watson died, she left the custody of her shack to Charlie Schmid. He took care of it as best he could.

Salvatore: Toward the end of her life she was such a determined woman. Never married. She was librarian, I think. But she loved the dunes more than any artist ever could. She knew every blade of grass. She became crippled.

Josephine: With arthritis.

Salvatore: But she just wanted to go out to the dunes. And Charlie [her neighbor] would carry her on his back up to the cottage every spring and every fall, and whenever she had to go into town. He took care of her. And for that, she left him her cottage.

Josephine: She dramatically died, crawling on all fours to get to her cottage. She could drive her jeep. But her cottage was in a rather difficult place.

Salvatore: It was on the top of a hill.

Josephine: So she had to get there by going up the hill. Peg Watson was another one who was totally devoted to the dunes, and who had nothing to do with the arts.

Grace Bessay

Josephine: Take Grace Bessay.

Salvatore: She had the longest lawsuit in the history of American law against the U.S. government. How many years was it?

Josephine: Oh, it was twenty years at least. Peter Clemons can inform you of that.

Salvatore: Out of her own pocket. She paid for everything out of her own pocket.

Josephine: She was completely, well almost, bankrupt by it.

Salvatore: That kind of passion. Jesus.

Josephine: That shows you what kind of passion these people have. She wanted so much just to keep her way of life and her cottage. They were giving her such a hard time because she couldn't have "improved property" status because she didn't observe the zoning code of the town. They said, "You don't have the plumbing, you don't have this and you don't have that." And she said, "For God's sake, if I did that you'd throw me out because I'm not maintaining historic property." All she wanted to do was to have her shack and let them leave her alone.

Laura and Stanley Fowler

Josephine: The Fowlers. They had nothing to do with the arts. Their place was beautiful. It was so beautifully kept. They came back every year. What they loved was to be with everybody else. They were kind of gossipy. They'd pick up little news here and there, you know. But they didn't bother anybody, and they were so loving and kind.

Harry Kemp

Josephine: Harry Kemp, of course, represents the dunes. He's the dunes. His spirit is out there still. We were dear friends of Harry Kemp's. He was diabetic. He had to take insulin everyday. He would leave the needle out there with the medicine. And so he would walk to the Backshore every day in good weather. We're talking about the spirit of the dunes, Harry Kemp. Some of his best poems were about the Backshore. Those poems will probably live. Harry lost recognition in his last years. But he was a highly successful writer in his time, and of course the last bohemian in the Village. He knew everybody. He was a good friend of O'Neill's. In a sense, he was sort of the bulkhead out there of the writers, starting in a sense with Eugene O'Neill, with Harry coming later.

Jeanne Chanel (Frenchie)

Salvatore: A remarkable woman. She chucked a promising career in theatre in New York and chose to come here and to build her own place with her own hands.

Josephine: Frenchie Chanel used to post the tern nests, long before the Park ever came.

Salvatore: She'd make a cordon of dune wood around the nesting areas of the birds. She did this on her own. She knew those little fellers were in danger.

Josephine: She'd take care of any wounded seagull. She had a remarkable sense of things.

Robert Gibbs

Salvatore: Our first superintendent was Gibbs, Robert Gibbs. He was so wonderful. He really got himself involved with the community. At that time, all of our sister towns on the Cape still looked upon the Park as an intruder into the way of life.

Josephine: Here it was considered a great intrusion. Here the common cry was, "Oh, you've stopped our development cold. The Park has taken everything." Now you can hardly find a single person in this town of any persuasion who doesn't believe the Park has saved it. And it's true.

Salvatore: Gibbs used to say, he had been to the Great Smokies. He was a chief ranger there. He said, the one thing we learned the hard way, when you go to an area that has a uniquely rich cultural background, you respect it. You let it lie, let it be, providing they don't transgress into your unlawful procedures, you know. He said, "Every now and then we'd be walking around [the Great Smokies] on our horses and we'd hear, 'Boom!' It would be another still blowing up. *[Laughter]* We'd just turn the other way." He tried to bring that philosophy here. For the long time he was here, he was much loved by the lower Cape people.

The Dune Shack People

Josephine: It's the people in these places [the shacks], the way of life they have in these places, that forms the cultural background of the thing. God knows everybody out there has been an extraordinary individual. We are all uniquely different from the average run-of-the-mill. We know everybody out there. Unfortunately, the case [to preserve the shacks] has been made around the constellation of artists and writers and so forth. But really it goes further than that. This is a community of unique individuals, unique in their character. They want isolation. They want solitude. They want independence. They do not want to be bothered by anybody. What you have to look at here is a way of life. This is what we are trying to save.

One main conclusion that may be drawn from life history sketches like these is that the dune shack society known by the Del Deos is extremely personal, and very self-aware. Long-term dune dwellers not only know about one another, but they love to talk about one another, and to draw life lessons from the stories. Even newcomers like Robert Gibbs, the first superintendent with the Seashore, can be brought in and preserved as a personality in the oral traditions of the dunes. From the 1920s through the 1960s, a distinct dune shack society grew to be a very personal community, rich with colorful people and outrageous stories, and more than full with a sense of itself.

Preservation and the Cape Cod National Seashore

The formation of the Cape Cod National Seashore by Congress in 1961 was a milestone in the history of the dune shacks. In the versions of local history told to me, dune dwellers supported the Park's creation at Provincetown. Dune shack residents related this part of history with considerable pride. Dune dwellers played essential political roles in getting the Park fully established in the Provincetown area, according to local accounts. Grassroots activists in Provincetown sought to preserve a threatened way of life on the lower cape by protecting the dunes through the Park. The perceived threat was from excessive land development for summer tourism and second homes, development controlled by outside interests primarily driven by short-sighted profit rather than the long-term benefit of Provincetown. Grassroots activists viewed the Park as a way to protect areas around Provincetown for their traditional values.

In the Provincetown area, the political fight focused on the fate of the "Province Lands," 3,000 acres of relatively open commons including shoreline, wetlands, and dunes, held by Massachusetts. In 1654, a chief of the Nausets sold the Province Lands to Governor Thomas Prince for use of the Plymouth Plantation Colony. In 1670, the Province Lands were among the first areas in America set aside as a fishing reserve by the General Court of the Old Massachusetts Bay Colony. The issue was how much of the Province Lands would be transferred to the Park, and how much would be retained by state, town, or private owners for future development. Grassroots preservationists wanted all the lands to go to the Park. They drew substantial support from the artists and writers at Provincetown, and eventually from a majority of the town's voters.

The fight to preserve the Province Lands for the Park was catalyzed by Ross Moffett (1888-1971), a painter and resident of Provincetown since 1913. Josephine Del Deo, a writer and dune shack resident, teamed up with Moffett in the campaign to organize public support to preserve the Province Lands. She chronicled the three-year political fight in her biography of Moffet (Del Deo 1994:317-345). Josephine and Salvatore Del Deo remembered this personally-difficult, grassroots political fight for the Park in our interview:

> Josephine: I wrote a biography about Ross with a chapter devoted to this. I was frightened that this history would go by and never get written down. I was writing about his life, so this was something that I couldn't miss. It was major. As you know, it was the first Park in the country that was placed inside an already-developed area. That's what made it so unique. And that's what made it so tough. We saved 1,500 acres of the Province Lands in our fight to establish the National Seashore. We had an original tract from the State of Massachusetts of 3,000 acres. The town fathers wanted to grab 1,500 acres of it for development. We fought that off. That was a remarkable fight. As a result, the whole of the Province Lands went into the Park.

> Salvatore: You can't imagine what an incredible battle this was. My poor wife. I'll tell you, she has done many things, but this has certainly got to be one of the major victories. She fought against the local newspaper, the selectmen, millionaires like Walter Chrysler, people who had lots of money, who had envisioned a tremendous development from Monomoy all the way to the Province Lands, wall-to-wall with development.
> Josephine: As soon as they knew the Park was coming in, it was like alligators.
> Salvatore: They started moving fast. And she had to defend her position on behalf of the little people on the back street, the Portuguese people, the innocents who had never been threatened before.

> Josephine: We have to say it was with my dear friend, Ross Moffett, the artist. We shared the burden. This man was tremendous. He was the only one from Provincetown in 1959 who went to the Senate hearings in Eastham concerning the Park. He was the only one who came to speak on behalf of the Park. And when he came back, he talked to me about it. I said, "Can I help you?" And that was the toughest thing I ever said because he said, "Yes," and it was the next three years. He was a remarkable man. We formed this

committee, the Emergency Committee for the Preservation of the Province Lands. So many people helped us. But Ross Moffett was the one.

Salvatore: Ross was later recognized by the Park as their first archeologist in residence. He knew more about the cape than anybody else. He walked every area of this place, from here to Monomoy. He had lived here since 1913.

The Emergency Committee for the Preservation of the Province Lands, organized in 1960, grew out of the Provincetown Property Owner's Protective Association, a group working to protect the traditional qualities of town. According to Del Deo (1994:321), its original list of members included more than fifty artists and writers in Provincetown. At the root of this movement was the preservation of local traditions against destructive forces of other traditions, in this case, large-scale real estate profiteering, as described by Del Deo (1994:322):

What we sought was seemingly unrealistic from the contemporary view: *non-acquisition* of land to develop; *non-satisfaction* of investment interests. Over and against this were the traditional forces of exploitation best represented by real estate investors and planners who argued, sometimes subtly but always determinedly, against us... It was the comfortable short view expressing itself in a society ever self-serving, no more so in Provincetown than in any other community.

The battle portrayed here was between two visions for the future of Provincetown and its surrounding open lands – a future with its traditional values, or a future without them. The determinant force was real estate development, its extent and type. The preservationists fought for placing all the open lands surrounding Provincetown into the new Park, keeping them out of the hands of land developers, to help preserve the traditional values of the town and its way of life. In their assessment, this was a far-sighted vision rather than the "short view."

As told by Del Deo, between 1959-61 the fate of the Province Lands was debated by Congress deliberating the Park bill in Washington, and by the public in local town meetings and federal hearings on the cape. The Emergency Committee for the Preservation of the Province Lands pressed for their vision at each venue, against the efforts of land developers who worked to retain a piece of the Province Lands for development. The tug-of-war for the public's sentiment at Provincetown was decided at the annual town meeting in 1961, where four alternative questions about the Cape Cod National Seashore were voted upon by the residents of Provincetown. Del Deo (1994:331) interpreted the outcome of Provincetown's final vote:

When the vote on this last question came, it was 144 in favor of, as opposed to 61 against, retaining the entirety of the Province Lands as they were for inclusion in the National Seashore Park. It was a moment in conservation history which had brought to fruition, as so few moments have, a resolute stand on the environment, an intelligent plan to carry it out and a final determination to see it through. No other town had so rallied to a view of the future. No other town had supported the Park so wholeheartedly. We were again unique on the Cape. The input of the artists and writers throughout this entire effort had been crucial.

The congressional bill establishing the Park passed in 1961. The local grassroots campaign in support of the Park continued through 1963, until the Province Lands were finally transferred from Massachusetts to the Park. Del Deo (1994:322-323) summarized the local effort:

The fact is that, eventually, Provincetown, of all the towns on the Cape scheduled for inclusion in the Seashore, was the only town to endorse the Cape Cod National Seashore Park unconditionally and to enter the Park exactly as the original first piece of legislation had suggested, without one acre being deleted from the proposed land mass. How did this happen? Certainly not by poetic pronouncements but through a carefully orchestrated and intense campaign which began in 1959 and which did not falter for the next three years.

From the point of view of Del Deo, achieving the Park was a personal, hard-fought victory. Artists, writers, and others in Provincetown who fought to preserve the town's traditional values and its distinctive way of life, believed that they had a personal stake in the new parklands. Having toiled in partnership with federal entities to create the Seashore, many entered this new era with high hopes.

Preservation Struggles, 1960-1985

Until the 1960s, dune shack society had no formal integrative organization, other than the informal ties between particular shack families. Dune shack society was a collectivity of neighbors living in family-owned shacks. Dune dwellers espoused values like autonomy, privacy, and personal freedom, and shared certain things, such as a passion for the dunes, historic roots, and certain common practices. There had never been a need to organize politically. There was no governing entity or representative association specific to the dunes. Their local municipal governments were either at Provincetown or Truro, where they paid property taxes depending upon the shack's location in the dune district. The dearth of political organization within dune shack society was to change after 1960. The changes occurred in response to perceptions of threats to the dune shacks and their traditional uses that materialized with the newly created Park.

With the formation of the Cape Cod National Seashore in 1961, conditions on the Backshore began to change markedly. The Seashore became the primary owner of the Backshore lands through the state's transfer of the Province Lands in 1962. The Seashore's enabling legislation authorized and defined procedures for it to acquire public and private land within the boundaries of the Seashore. During the 1960s, the Seashore proceeded to acquire land within its boundaries. The procedure involved identifying land and property owners and, depending on the status of the specific lot or structure, making an offer to a willing seller, or initiating condemnation procedures. Individuals with valid ownership to improved property (defined as buildings in place before September 1, 1959 with electricity, water, and sewer) were permitted under law to maintain ownership if they did not wish to sell. Somewhat over 550 structures were found to qualify as private property inside the Seashore. Most were single-family homes with titles and facilities that were deemed to meet the standards. The structures were allowed within the Seashore as privately-owned dwellings, adjoining or surrounded by park land. Some structures were occupied year-round, but most were second homes used primarily as summer residences. Other structures did not qualify. Somewhat more than a hundred properties were put under use and occupancy reservations throughout the Seashore, most for being post-1959 construction.

The Seashore's acquisition of property inside the Park directly affected the dune shacks. The Seashore asserted that the dune shacks did not qualify for private ownership within the Park, and the Seashore moved to acquire them from their owners. The dune shack residents asserted that their shacks did qualify. The facts of the individual cases proved complicated. Most shack owners had deeds of transfer for the shacks acquired through inheritance, gift, or purchase, but most shack owners did not have clear titles to the dunes beneath the shacks which were held either by other private parties or by Massachusetts, depending on the shack. Further, because they were rustic structures adapted to dune conditions, many dune shacks had no (or non-standard) systems for electricity, water, and sewer, raising questions as to their improved property status.

Most dune shack residents were not willing to relinquish title to their homes to the federal government. Residents understood that once the federal government held title to the shacks, the government might compel families to leave. Then the shacks might be razed. In their recollections of this historic period, Mildred and Maia Champlin recalled Herb Olson, the superintendent of the Seashore, telling the family that his goal was shack removal:

> Mildred Champlin: Olson stood on the beach and he told us. We were standing on the beach, and he said to us, "I am looking forward to the day when all of these are gone and it's back to nature."
> Maia Champlin: And he's talking to homeowners. What if somebody came up to you in your home, that you've worked on for decades, and said, "I can't wait to see you gone?"

To dune dwellers, the federal government's ownership of the shacks presaged the end of the dune shacks and the society of dune dwellers with their traditional patterns of living.

In an effort to preserve the dune shacks and their traditional uses, dune dwellers organized politically by forming a representative association. For the first time, the dune shack families were taking collective political action as a united body. Dune shack owners created the Great Beach Cottage Home Owners Association in 1962. Their purpose was to work as a united front to prevent the federal acquisition of the cottages on the Backshore. In the organization's name, the term, "cottage," was chosen over "shack," as the latter term had a connotation of "unimproved property," a legal category of structure subject to taking by the Seashore.

A news article entitled "Cape Cod Beach Owners Have a Problem" in the *Cape Cod Standard-Times* (October 14, 1962) covered the formation and purpose of the new organization of dune shack residents:

> On their part, the "dune dwellers," usually an offish lot who eschew neighboriless in search for aloneness, have banded themselves into a united, determined group already and, one gets the impression, eager to fight for their rights to live in isolation. Normally free-thinkers, they have formed the Great Beach Cottage Owners Association with the intent of raising their voice in high circles, though it is an anomaly. Heretofore, Back Shore cottage owners have sought only freedom to live, paint, work, or write as they please without interruption, and if their nearest neighbor is two miles away – so much the better… "A place in the dunes," said Ray Wells at a recent meeting, "is a way of life."

The article ran with a picture of seven members at Ray Wells' home, including Ray and Nicholas Wells, Grace Bessay, Hazel Hawthorne Werner, and David, Constance, and John Armstrong. According to the article, the purpose of the organization was to preserve a "way of life" on the dunes, including "freedom to live, paint, work, or write." It also identified the apparent contradiction of united group action by an "offish lot" that customarily did their free-thinking autonomously, "in isolation." The organization planned to appeal to "high circles," including Congress and the Park Service, to find ways to retain their cottages and preserve their uses.

A membership list of the Great Beach Cottage Owners Association circa 1962 identified 40 members and one "Counselor at Law" who "contributed time and advice to the organization, and represents some of the members individually." The group was an association of "cottage owners," persons with legal interests in the dwellings. Other shack residents and users, a substantial set of people, were not listed. The membership list is summarized in the following table, linking members with their cottages (as of 1962), as well the cottage's current identity (as of 2004, based on information obtained in this study). As shown in the table, all dune shacks had an owner on this membership list except one, the Malkin-Ofsevit-Jackson shack. At the time of this study (2004), at least fifteen of the shacks were still being used to some extent by the cottage owner's families or their designated heirs-caretakers (two Adams shacks, Champlin shack, Malicoat-Lord shack, Gelb-Margo-Zimiles shack, two Werner shacks, Fowler shack, Clemons-Benson shack, Tasha shack, Schnell-Del Deo shack, Schuster shack, Wells shack, Jackson shack, and Armstrong shack).   In 2004, four shacks had occupants not directly connected to the owners in 1962 (C-Scape shack, Beebe-Simon shack, Isaacson-Schecter shack, and Dunn shack). In 2004, six shacks in the eastern group of shacks were gone, removed by the Seashore (Schmid shack, Bessay shack, two Fuller shacks, Stanard shack, and Vevers-Pfeiffer-Geise shack). And one additional shack was gone, destroyed by an accidental fire and never rebuilt (the Ford shack).

Members of the Great Beach Cottage Owners Association, Circa 1962

Members (Names and Number)		Member's Cottage	Cluster	Current Shack
3	Mr. and Mrs. Don Burns; Jean (Cohen) Burns	Burns-Cohen	Western	C-Scape
1	Leo Fleurant	Fleurant	Western	Beebe-Simon
1	David Adams	Adams - I and II	Western	Adams - I and II
1	Nathaniel Champlin	Champlin	Western	Champlin
2	Philip and Barbara Malicoat	Malicoat	Western	Malicoat
2	Jan Gelb and Boris Margo	Gelb-Margo	Central	Gelb-Margo-Zimiles
1	Hazel Hawthorne Werner	Werner (Euphoria)	Central	Euphoria (Werner)
		Werner (Thalassa)	Central	Thalassa (Werner)
2	Stanley and Laura Fowler	Fowler	Central	Fowler
2	Al and Doey Fearing	Fearing	Central	Clemons-Benson
1	Rose Savage Tasha	Tasha	Central	Tasha
1	Jeanne Chanel	Chanel	Central	Schnell-Del Deo
2	Theodore and Eunice Braaten	Braaten	Central	Schuster
1	Margaret Watson	Watson	Central	Isaacson-Schecter
2	Nicholas and Ray Wells	Wells	Central	Wells
2	Mr. and Mrs. Randolph Jones	Jones	Central	Dunn
3	Josephine Ford	Ford	Central	(Gone)
	(No member listed)	Malkin-Ofsevit-Jackson	Central	Jackson
3	David and Constance Armstrong; John Armstrong	Armstrong	Eastern	Armstrong
1	Charles Schmid	Schmid	Eastern	(Gone)
1	Grace Bessay	Bessay (Red Shack)	Eastern	(Gone)
1	Andrew D. Fuller	Fuller (Concrete Shack)	Eastern	(Gone)
		Fuller (Joe Oliver's)	Eastern	(Gone)
2	Mr. and Mrs. H. Stanard	Stanard (New Yorker)	Eastern	(Gone)
4	Mr. and Mrs. Anthony M. Vevers; Mr. and Mrs. Chet D. Pfeiffer	Vevers-Pfeiffer-Giese	Eastern	(Gone)
1	Francis Willemain	?	?	?
40	Total Listed Members			

The owners' united efforts to resist the federal taking of the dune cottages placed the dune dwellers and the Seashore in an adversarial relationship for the first time. As discussed above, many dune dwellers previously had lobbied to create the Park. They had done this in an effort to preserve traditional ways of life at Provincetown against excessive real estate development. But when it appeared to the dune dwellers that the preservation of the dune shacks and their traditional uses was not part of the Park's vision, these early cooperative relationships soured, and then turned adversarial in court.

As a united effort, the Great Beach Cottage Owners Association was active somewhat less than ten years. As things developed, the federal acquisitions of the dune shacks were determined individually, rather than as a class action. Individually, some shack owners retained legal assistance to represent their interests. For some this entailed long and expensive litigation. The longest was litigation by Grace Bessay, which lasted from 1967 to 1991, reputed to be the longest civil suit ever against the federal government, according to local oral tradition. Other shack owners with more limited means did not litigate for long. As cases were settled, the association lost members. Eventually, all cases were decided. The ownerships of all dune shacks except one (the Malicoat shack) were taken by the federal government. The previous owners lost legal title to the shacks.

The dune dwellers were correct in their understanding that the Seashore intended to legally terminate the residencies of dune shack families. The Seashore set termination dates for each family through "reservations of use." In the majority of cases, under settlements between the owners and the federal government, the Park issued either a transferable 25-year use (a time estate) or a non-transferable lifetime use based on certain occupants (a life estate). During this period, the persons holding the reservation were legally permitted to determine who used the shack, within specified parameters. When the terms of each reservation ended, the legal right to use the shack ended. In sum, the reservations of use established finite legal tenures for dune shack families. But the settlements did not say whether or not families using the shacks would be evicted at the end of their tenures, nor did they say whether or not the dune shacks would be removed. Under these new stipulations, dune shack residents continued to use their shacks, which were now legally owned by the federal government, except for several shacks in the eastern cluster that were removed by the Seashore.

Preservation Struggles, Post 1985

The struggle by dune dwellers to legally preserve the dune shacks and their traditional uses saw a revival after 1985. This second effort was spearheaded by leaders of a newly-formed nonprofit organization called the Peaked Hill Trust. The grassroots effort eventually led to the recognition by state and federal agencies that the dune shacks had values of an historic nature. According to local accounts, the second preservation campaign was catalyzed by an action of the Seashore – the razing of the Schmid shack.

Charlie Schmid, also known as "Dune Charlie," was a respected eccentric within dune shack society, a full-time dune resident whose long-term studies of tree swallows on the Backshore dunes had gained international recognition. Schmid's multi-storied shack was unique on the dunes and considered a local landmark. Phil Malicoat and Anne Lord described the shack's construction:

Malicoat: I was a particularly a good friend of Charlie Schmid. I helped him build a good deal of his shack. There's two structures there. There's one that was originally built by a man named Meads. He sold it to Charlie for a hundred dollars back in like, 1957 or 1958. That shack eventually became buried. But what he did was, he wanted to build something on top of it. I helped him build something on top. Now he had this place he could go underground for the winter. He just had one little burner, a kerosene burner that would warm that place right up, no matter what the weather was. When they bulldozed that shack they probably didn't realize this. I don't think they dug it up. There's another shack underneath it that could probably be uncovered. *[Laughs.]* He was wry person, like Harry Kemp. And you've heard all about the birds?

Lord: He went to conferences to consort with Konrad Lorenz. He met him, taking these interesting data to meetings.

Malicoat: To Switzerland.

A colorful description of the shack was presented in Josephine Del Deo's *Compass Grass Anthology, A Collection of Provincetown Portraits* (Del Deo 1983:43-45). Her affection toward Schmid and his multi-storied shack is evident in descriptions of a visit she paid to his place with her husband:

> By any natural law, the place should have fallen into the sea long before. It was a driftwood "Lego" set no manufacturer had ever produced, a construction that defied gravity. As Charlie careened carefully along the rotting planks of the top level of his crazily cantilevered porch, God, in some kind of tacit arrangement with the odds, always tipped his balancing act in favor of the return trip. Down below, another deck shot out, as Wright would have said, "to fit the contours of the land.".... [In] that surrealistic interior, we became confused as to which level we were sitting in, but the concern for our whereabouts gradually assumed the casualness of Charlie's comings and goings between decks as he primed the pump, frequented one of three bathrooms or retrieved a beach treasure for us to admire. We sometimes thought of the thirty-foot drop out back, but decided that if we leaned to the north'ard, we might escape Newton's Law. Balzac lay curled unconcernedly by the stove as Charlie jammed in another piece of driftwood. The clatter of the lid was a sharp sound against the monotonous pulse of the sea outside. (Del Deo 1983: 43, 45)

Schmid received a life term reservation of use of his shack. He lived there full time, conducting bird studies at an adjoining wetlands. Over time, Schmid's health failed and he was nursed by friends away from the shack. He died in 1984. One week after his death, his shack was bulldozed. The event shocked dune shack society, according to many I interviewed. The incident became an iconic story in local oral traditions. It was told to me in various versions by a number of people. One person said the shack was bulldozed with Schmid's possessions still inside, the wreckage left scattered across the dune ridge for years, "like Appalachia." "Charlie was a very private person," said another dune dweller, who helped Schmid with his bird observations, yet "all his stuff was strewn out for everyone to see."

Dune dwellers concluded from this action that the Seashore intended to quickly demolish dune shacks when their reservations of use ended. This belief energized dune shack users to organize to save other shacks from similar fates. Julie Schecter, former director of Peaked Hill Trust, described the leveling of the Schmid shack, and the efforts that followed to preserve the others:

> The shacks were under leases from the Park to individuals, some of them under life leases, some of them under timed leases. One of the life leases ended, which is to say, the man [Charlie Schmid] died. It had been a while, a stretch of years, since one of the leases had expired. There were a number of us who were convinced that park policy regarding the shacks had moderated a bit, and the shacks that were slated to be removed from the Park would not be removed from the Park. Well, this particular fellow's shack was bulldozed. And quite a few of us said, "Oh, my!" They did exactly what they said they were going to do, surprise, surprise.

The superintendent of the National Seashore told me completely flat out that they wanted to get rid of the dune shacks. They had to certify that the shacks had no historic value, and then they could remove them. He in fact did hire somebody to look at the dune shacks, in the dead of the winter when nobody was out there, and take pictures of them, and got an attestation saying the shacks had no historic value. I didn't actually believe that the most interesting thing about the shacks was that they were historic. I don't really know too many people who thought that's why they were interesting or a resource. But the guy who had control over this had just told me that either they're historic or they're cinders.

So we went around and asked as many people as we could for as much information as we could about the history of the shacks. Who had been there? What had gone on? We contacted conservation organizations, historic organizations, all the selectmen in the pertinent towns. Hazel Werner had been a part of the Eugene O'Neill-John Reed group and she had a lot of contacts. We used all of those contacts. We wrote letters to people and called them up and said, "What do you know? This is the situation we're in. Is there anything you can do to help us out?" We were really lucky that we had a letter from Jack Kerouac saying that he had written part of *On The Road* there. We got a number of letters from various people. One of the people who went out there often was Joe Hawthorne, the conductor of one of the Ohio symphonies. He went out every year until he became disabled. There was an article in the paper. People called me and said, "What can I do, what can I do?" I said, "Write to the Massachusetts Historic Commission."

There was a meeting that the Massachusetts Historic Commission held right here in Provincetown. Something like 200 people showed up to this meeting. It was stunning, just stunning, and all this over shacks that most of these people had never visited. But it was part of their community, part of their world. On a number of instances since then, I had people come up to me, and say, "You know, I looked at those shacks for a decade or more, and I knew that they were there, and I just really liked that they were there. I didn't think I could go visit them. It was just that they were there." That gives me the chills. They're really special to a lot of people.

Initially the Cape Cod National Seashore found that the shacks were not eligible for listing. By the procedure they follow, the finding goes to the Massachusetts Historic Commission. By and large the Massachusetts Historic Commission has no reason to disagree with the local park service representatives. In this particular instance, they received more comments on this issue than they had received on any issue they ever dealt with in their history.

This grassroots action ultimately led to a finding by the Massachusetts Historic Commission that the dune shacks were eligible for listing on the National Register of Historic Places. The eligibility status meant that other dune shacks probably would not be immediately removed at the termination of a lease or reservation. The eligibility finding also identified the shacks' historic values as something to be considered in the shack's management, maintenance, and authorized uses. The eligibility finding was pivotal in changing institutional assessments of the shacks, according to Julie Schecter:

The shacks are now seen as having some value, that they are a resource, which originally they weren't. They weren't a resource for them, so they wanted nothing to do with them.

The eligibility finding also affected assessments of the roles of dune shack residents. Under this new perspective, shack residents who appropriately maintained and used shacks might be contributing in preserving something with historic values. They were not simply occupants marking time in a structure scheduled for eventual demolition.

Historic and Cultural Values

The Dune Shacks of the Peaked Hill Bars Historic District was determined to be eligible for the National Register of Historic Places in 1989. While eligible, the shacks had not been listed in the National Register at the time of this study (2005) because administrative steps had not been completed. The historic district comprised an area of about 1,500 acres covering the viewsheds of the dune shacks along a two-mile stretch of dune ridge. The historical association of the dune shacks with Provincetown writers and artists was a principal factor in the finding of eligibility as a historic district, since these persons were involved with the development of American art, literature, and theater. A second factor of eligibility listed in the finding was the physical structure of the shacks themselves, which were said to represent "a rare fragile property type." Third, the shacks and the dune together were said to represent a historic cultural landscape "comprised of a distinctive, significant concentration of natural and cultural resources united by their shared historic use as a summer retreat for the Provincetown colony of artists, writers, poets, actors, and others." One shack in the district (the Tasha shack) was found to be significant for historic associations with the life of an individual poet, Harry Kemp, but other direct associations of shacks with specific individuals were not identified. The determination of eligibility recognized that the dunes and the dune shacks, in representing a pattern of historical use and cultural symbolism, "were frequented not by isolated individuals, but rather by a collection of varied artists united by the dynamic process of artistic creation."

The finding of eligibility identified the "dune landscape" as the "lynchpin of the district's cultural importance." The dunes were said to be "the source of natural beauty and artistic inspiration," and they "provide the key to the district's existence and to its fragile historic character. The shacks are the surviving material artifacts that convey the significant, continued historic use of the dunes' seaside setting over time." Further, "the dune shacks provided shelter while minimally intruding into the contemplative solitude of the environment that provided the impetus to an abundance of artistic and literary work. The shacks' unpretentious, predominantly one-room structure, their simple materials and craftsmanship, their mobility, and their lack of amenities such as electricity and running water enabled their inhabitants to experience a survivalist relationship with nature." The "period of significance" was identified as from the 1890s until 1960, the year the Seashore was established.

Subsequent to the Schmid shack action, reservations of use terminated for a number of other shacks, including the Burns-Cohen shack, two Werner shacks, Gelb-Margo-Zimiles shack, Braaten shack, Watson shack, and Jones shack. The Seashore did not remove them, though several remained unoccupied, experiencing substantial deterioration. Eventually, the Seashore arranged for all the shacks' continued use and maintenance in partnership with selected families or private entities. The Seashore put three shacks under the care of non-profit organizations (Peaked Hill Trust, Provincetown Community Compact, and OCARC) with provisions for access that included artist-in-residence programs. Such programs represented substantially different patterns from historic uses, as discussed in later chapters. By 2005, through the scheduled expiration of existing reservations of use, ten shacks would come under the legal management of the Seashore, with others following suit in upcoming years. With so many shack reservations ending, the Seashore sought guidance regarding the future of the historic district from the Cape Cod National Seashore Advisory Commission, an advisory body with representatives of six towns, the county, the state, and the Department of the Interior. A dune shack subcommittee created by the Commission periodically convened after 1990 to assist this consultative process, addressing questions such as, what should become of the district, the shacks, and the shack uses, and how might the values of the district be preserved?

In 2003, in response to a subcommittee report, the Provincetown Board of Selectmen raised a question as to whether the historic district was "traditional cultural property" for the dune shack residents, who might qualify as "traditionally associated peoples." In a letter to the Commission dated May 14, 2003, the selectmen stated that "some, perhaps all, dune shack dwellers are a protected cultural resource, and the Cape Cod National Seashore (CCNS) has legal obligations to manage this cultural resource in the way that best maintains, perpetuates, and strengthens this cultural group's continued access to and use of the dune shacks in which they live." The selectmen referenced developments in cultural resource laws and policies, including the *Guidelines for Evaluating Traditional Cultural Properties* (National Register Bulletin No. 38). If so, then an additional value might be considered in the Seashore's vision for the future uses of the historic district, its value as traditional cultural property. The query from Provincetown raised a need for additional ethnographic information on the customary and traditional patterns of use of the dune district. My current study was designed to collect and present that information.

As stated at this chapter's beginning, history is more than just a chronicle of milestones for those who have lived it. For many dune shack residents, the local history was deeply personal. Every shack resident I interviewed offered heartfelt stories related to the efforts to continue the traditions of the dune shacks. Like stories about their neighbors, the stories of struggle to preserve the shacks have become part of local oral traditions on the dunes. The historic narrative provided by Paul Tasha, presented below as an example, illustrates its personal nature. The Tasha family's reservation of use was scheduled to end the year of this study (2005). In our interview, Paul Tasha told his version of history, beginning with Harry Kemp's bequeath of the shack to his family, and covering the stipulations with the new Park, the struggle to preserve the shacks, and the ambiguities of the present time with its uncertain future. As a very personal ending to this historic chapter, Tasha's account is touched with sadness, perplexity, frustration, humor, and hope:

> The way it read, when Harry [Kemp] gave it to us, it said, "I give my dune shack to Mrs. Herman Tasha and her family." This is the formal name, my father being Herman. So I guess the shack belongs to the family, at the very least to the four kids that were alive then, and I suppose technically, the term "family" would include the grandchildren by now, I would hope.
>
> We didn't have a real title. You didn't need one back then. It wasn't the traditional way of doing things, not to belabor "traditional," but it just wasn't. You didn't need to do that. It wasn't necessary. And now, oh boy, now, you better have everything lawyerized. Had we taken the time before the Park came in, we had the opportunity. You just never thought you'd need it. We could have had it done. No one would have objected, because nobody really wanted anything out there [on the dunes]. It was only very eccentric weirdoes who wanted to be out there. "What the hell do you want to go out there for? There's nothing out there. The shacks don't even have plumbing!" It wasn't very popular.
>
> But they offered us a contract, "Either get out now, or take the 25 years or whatever, because you don't have a legal title." I said, "The way it was done in the old days, Frank Hendersen gave it to Harry, Harry gave it to us, that's how it was done. That's not good enough anymore?" Nope. The guy told me, "You don't have a legal deed and title, you don't have a leg to stand on. So basically you got to get out now, or take your 25-year deal." We hired two lawyers. They came back to us and said, "You'll never win. Take the deal. Take the

twenty-five and see if you can work it out in the long run, because there'll be new administrations, maybe more reasonable heads."

We had at one point asked the Park to give us life tenancy, for those of us that were alive, because they were giving life tenancy to some people. But they wouldn't do it. The guy told me, "Frankly, you're apt to live too long and we'd like to have this all over with within 25 years, so we'll give you 25 years. Or, you can prove you own it legally with a deed and a title and we'll leave you alone. But if all you have is that little piece of paper from that drunk poet, then you don't have a leg to stand on. So you can leave now, if you don't want to take the 25-year deal."

So, of course you take the deal, because it's either that or you get out now. At least you got 25 years. You kinda felt like you had gun to your head. 'Yeah, yeah, I'm a witch, shoot me.'

Most of the time, I don't know why they do what they do. They don't even know why they're taking the shacks away from us. They don't actually have a reason. There was a reason, originally. The Park came in and said, "We don't want to change anything. We want to preserve it." That was in the 1960s. The middle seventies we got a letter, a new administration. They want it pristine. They want to bulldoze all the shacks and create a pristine environment. And they did bulldoze, like three. Then there was an outcry and the National Historical Register, saying, you can't do that anymore. Their reason was to create a pristine environment. Okay, I can kind of understand that. But since now they can't, the only one reason they had for taking the shacks away was gone. So now you say, "Why are you doing this? You're just going to take it from us and give it somebody else because you have to keep them, mandated to make sure they exist. So, you lost your reason. So why don't you lose the whole damn idea?" *[Gentle laugh]* It's like, you know, kinda like trying to fight elephants with farts. *[Laughs]* You get steamrolled, you know? But, you ask at this stage, "Why are you doing this?"

Definitely there's a better attitude now. But still, nobody is willing to say, "You know what, this was just a bad idea, we should have never done it. We did it with a pen, so let's just turn the pen over and use that rubber thing on the other end and put it back like it was supposed to be when Kennedy came in, when nobody was supposed to lose anything."

We were all basically promised nothing would be taken away. So you feel betrayed, you know. It may not be fair for me to compare myself to a Native American, but as far as I'm concerned, what they're doing is no different. You make a promise, then you change your mind and say, "We're going to do this." How can they do that? It's mind boggling that they feel so free to destroy people's ways of life without thinking that it's a big deal. It is a big deal. It's a huge deal.

I've gotten to the point now, and I hate to say it, but for the last couple of years I go to the shack and mostly I wind up sad and melancholy. It's like when my mother was really old and I knew she was going to die pretty soon. I didn't know when. But I knew it wasn't long. You just had that feeling, like, 'Well, I get to see her again, but I won't have many more times.' It's like that with the shack now. I get to be here again, and as wonderful as it is, now it's got that dark cloud over it of loss, of impending loss, you know. You can't help but while you're there, you think, 'How wonderful it is to be there and how lucky you are,' and then you remember, 'Oh yeah, well, they're probably going to take it away.' And you can't figure out why.

Chapter 3. Dune Shack Society

The last chapter traced the history of the dune shacks, how they have housed coastguards and their guests, summer writers, visual artists, year-round Provincetown residents, seasonal families, and bona fide eccentrics. Such a disparate fellowship forged a distinctive pattern of living on the backshore edge of Cape Cod, one that has continued into the twenty-first century.

This chapter begins the description of contemporary patterns in the dune district whose roots reach into that past. The chapter describes the numbers and types of people residing in the dune shacks at the time of this research, and how they were socially organized. As will be shown, they were members of extended families, networks of friends, members of nonprofit groups, awarded occupants, renters, drop-in strangers, among others. Together, these kinds of residents comprised the people of the dune shacks as I observed them in 2004-05. The core of this society of users was a set of long-term residents, primarily but not exclusively extended family groups. They maintained the shacks, managed their integrity within the fluid landscape, and reconstructed them after catastrophes, typically with a passion and commitment toward preserving the shacks and their traditional uses. But the dune shack residents extended beyond this core. Substantial numbers of short-term users gained access to the dune district through the longer-term dune dwellers. Long-term residents typically allowed their shacks to be used by friends. Other users operated as nonprofit organizations through which access was allotted by more formal rules. Over time, some of these short-term users themselves became part of the core. The size of the population of dune residents, and the kinds of the dune shack users, are covered below. Further descriptions of the social organization of this population are provided in the next chapter.

Discovering Shack Society

The residents of the dune shacks formed an unusual society, a nonstandard, primarily-seasonal settlement stretched along barrier dunes on the Backshore. Being unconventional, "bohemian" in the older parlance, was a source of some considerable pride within the set of people I interviewed. They knew they comprised a distinctive community, far from run-of-the-mill.

How to go about identifying this society of dune residents presented initial challenges to me, a newcomer to the area. Who were the dune shack people? On the Backshore, the society seemed almost invisible, at least to an untutored eye like mine. Walking among the dune shacks was not like walking along many public beaches on Cape Cod where I encountered throngs of summer beach users baking in the sun on towels, reading paperbacks under umbrellas, running through the waves, building sand castles, and other shore activities. By contrast, the Backshore dunes were nearly always empty of people. When I walked out into the dunes and among the shacks during the day at the height of the summer season, I never encountered more than a few other people. The dunes offered relatively open, empty solitude to someone on foot. Now and then, a dune taxi tour passed along the jeep trails, filled with tourists peering from cab windows at the sights. The occasional person I saw, or small groups of people, were commonly at a distance, usually moving in some other direction. Were they dune dwellers? Or were they vacationers like those in the dune taxis, simply exploring the Backshore on foot? Knocking on shack doors was not a preferred strategy to find out, I was told. It was rude. A polite person didn't do that. It disturbed the privacy and solitude of residents, glibly called "shackies" or "dunies" by some speakers. This local vernacular spoke of social boundaries demarcating *insiders*, those who lived on the dunes, from *outsiders*, people like me. The impropriety of knocking on shack doors clearly spoke of local etiquettes, that is, customary practices of a social group. But whose customs were these? How many were there? And on what basis were they an identifiable group?

I received help in discovering shack society at the Flying Fish Café in Wellfleet, the second town south from Provincetown. Emily Beebe took an hour off work to meet with me there for an interview. She herself was a relatively new member of the dune shack society, the "north neighborhood" of shacks as she explained. She and her friend, Evelyn Simon, successfully bid on the Leo Fleurant shack in 1993, becoming residents under a long-term lease with the Park Service. Before that, she and Simon had not been part of the dune shack society. During the interview, Beebe recounted her own experiences as an outsider looking in. Her initial encounters with dune shack society helped me to understand its basis, how to become a dune dweller, and as an anthroplogist, how to socially construct a hard-to-see group. I was surprised to learn that Beebe had never stayed at a dune shack, prior to winning the bid for her shack.

Wolfe: You had stayed in other dune shacks before you got this one?
Beebe: Never. Never.
Wolfe: What was your relationship with the dunes before your proposal?
Beebe: I used to walk out there all the time, just walk around the shacks, sit on the decks, take photographs, and just meditate, be out there. I had actually been out in 1988 or 1989, when I worked for the town of Provincetown... David [Adams] had to address the septic system for his place [on the dunes]. So he asked me what he needed to do to be up to standards. I said, "I think you need to put in a Title 5 septic system." He said, "You're kidding." And I said, "If the Park wants you to have a Cadillac, that's what you need to do." So that was my first introduction to this neighborhood. And I remember distinctively going out and sitting out on the bench [on his deck] and looking over at this place and wondering, what's going on over there? Why is that place all boarded up? What's happening? Really wanting to engage with David about it, but not wanting to invade his privacy. It was really trippy when you think of it in retrospect, that a few short years later this was our charge, to take care of it and to fix... Incredible.
Wolfe: So that was your connection to the dunes before, walking, and then seeing the shacks. And how about Evelyn?
Beebe: Same, the same. We liked to drive on the beach. We used to drive on the beach a lot. You know. She had always admired them. The shacks – they're like old people who you admire, who you want to talk to sometime, because you know that it would be such an interesting conversation, but you don't want to invade their space. That's what they feel like. At least to me, they do. So you admire things from a distance, these places. And we did that for years. I think that they kind of have that [effect] for people.
Wolfe: And they are unique for you, compared to other places on Cape Cod? There are a lot of cottages on Cape Cod.
Beebe: Yes. Absolutely. Because they're out there by themselves. They survive. And they more than survive. They're out there watching the ocean everyday, everyday, just collecting all this energy. They're little churches, each one of them. It's just, phenomenal. I think you can feel the magic of different buildings anyway, being inside them, being around them.

In this retrospective account, Beebe and Simon were users of the dunes as visitors, exploring, photographing, meditating, and the like. Beebe occasionally even experienced the shacks, walking among them, admiring them, and briefly sitting on decks. But Beebe clearly felt like an outsider to the social world connected to the shacks. She imagined the wonderful histories connected to them, but did not know how to start the conversation. She was an outsider looking in. But soon after winning the bid, an unusual thing happened to Beebe and Simon's social world. Their social world "exploded." The lease making them the residents of what had been Leo Fleurant's shack opened a door to the dune shack society.

According to Beebe, substantial numbers of other people were eager to be part of rehabilitating her shack. Unsolicited offers of help came to her and Simon, from people in and out of Provincetown, and from other shack residents, as described by Beebe:

> The process of rebuilding the shack created a whole other community for us. When we got this place in 1993, the place was falling apart. Literally, you'd walk in the door and you'd fall through the floor. One of the things that happened was, all these people started calling us, writing to us, being in touch with us, people who we had not known or had much contact with. We had people who just wanted to be part of that. We would meet every Sunday to go out to the shack when we first got it. We never knew who was going to show up at the parking lot at Race Point to come out with us. Now we have all these friends of ours who come out to use it on a regular basis. We have, basically, built a whole other family that comes out to the shack and uses it. Our friend Lee, our friends Kay and Larry who live in Brewster, and friends of theirs, and that sort of thing. So it's really an amazing thing that happened. This little gift exploded our world – it made it into so much more. It was really cool.

The volunteer help was not organized through any formal group such as the Peaked Hill Trust or a Provincetown art association, but through informal word-of-mouth. Beebe and Simon became beneficiaries of an informal, long-standing community of people connected to the shacks, or people wanting to be connected. They were experiencing a tradition of mutual aid among shack lovers. Also, they were falling into a customary practice of reciprocity, opening their shacks to people who contributed labor in their care.

A second type of assistance soon followed. Heads of three neighboring shacks, Nathaniel and Mildred Champlin and David and Marcia Adams, quickly established connections with the newcomers. They specifically came over to instruct them in the oral history of the dune shack residents and the ways of successfully living on the dunes. They gave them copies of old photographs, now part of Beebe's album documenting the history of the Fleurant's shack. They also offered assistance in practical exigencies, such as procuring water and getting cars unstuck from sand. Beebe recounts how these lessons from her dune neighbors began:

> Beebe: When we started with our project we had three or four months of working on it before our neighbors came out [to their own shacks], before we had any contact with them. We started in October and November. The Champlins come out every month and start their generator. They'll come out and heat up a can of soup in the wintertime, sit and do their thing, they're such a trip. They're great. And we just kept missing them when they would come out. The Adamses didn't come out until June. But as soon as they came out and we managed to coordinate, the Champlins came over. They gave me photographs. They told me stories about Leo. They told me the history of the building. They told me things about what we need to do with water. They told me about the best way to drive on the road out there because there's elevated ground water out there and you need to figure all that stuff out. We've pulled each other out of the sand I can't tell you how many times…
>
> Wolfe: You're saying that the assistance and the information were given gratis to you?
>
> Beebe: Oh my God, yes. And, "p.s., come on over for a glass of wine when you're done doing whatever you're doing."

Wolfe: And you didn't mind them? You weren't out there to be private?

Beebe: No. Absolutely not. They give themselves and their knowledge and their understanding and their experience. And it's such a gift, because it's fifty years of history. And it's [about] Leo, the fellow that I'll never know but yet I know, because of him being alive through them. And as I got to learn the building [by rehabilitating it], all the stories they were telling me started coming true in a certain way. I know every single board in that building because we really had to take it all apart and put it back together in a certain way. It's really amazing. They're just so great. They're very generous folks. And, you know, we share the same passion and understanding about how to live out there. I think that that's really significant. Totally neat folks.

A third type of unsolicited instruction arrived as well, information about the legal and regulatory histories of the shacks. The materials came primarily from Josephine del Deo, who also offered advice about dealing with federal government entities, according to Beebe:

All these people started calling us… One of them was Josephine Del Deo. She's still very much around. Her association is with Frenchie's shack. She lives in Provincetown. She is awesome. I would call her the head of this whole movement to save the dune shacks, right behind Barbara Mayo. She was the head of the dune shack subcommittee. She generated hundreds of pages of documentation on the shacks themselves. She did a lot of footwork. She carried the standard for the places for years. As soon as we got our place, she just buried us with paperwork, saying to us, "This is what you have. This is what you need to protect. You need to understand the National Park Service so you can deal with that." So she gave us a lot of background information. It has been absolutely invaluable.

Del Deo provided details on the sociopolitical contexts for the actions of dune residents and the Seashore. According to Beebe, certain Seashore documents suggested that there might be no people of historic significance associated with their shack. She said she felt "devastated" when she read that. After months of work on the Fleurant place, after learning about the life of the "little old man" who had spent twenty years of his life out there, like a hermit, to read someone calling him "insignificant" was "absolutely devastating." It started her on a "search for the truth," for what it is that makes the shacks significant. From this information, Beebe perceived that her family was now playing an important part of a longer, dynamic history. They were entrusted with a role in preserving the shacks and the ways of living connected with the shacks.

I could see that after this induction, Beebe and Simon no longer were dune users standing on the outside of dune shack society, looking in and wondering, like me. They were now part of that social group. Beebe stated that she was now part of a community, a new "family" that uses her shack, a social group that provides mutual support and that shares certain things in common:

We help each other in so many ways, because we have the same needs, we have the same parameters that we're living within. But I think the overall characteristic is that these are people who understand how to live out there. It's not about changing the way to live out there. It's changing the way you live to live out there, so that you're adapting to what is presented to you, which I think is the biggest piece about being out there anyway. You're not imposing your own will and your own way. It's not going to work. Your will, your way, your lifestyle is gone. And you have to come in and step in to this already-established mail shirt or whatever it is, so can deal with whatever is going to come up when you're out there. Whether it's working on your place, which everybody has to do, in a 45-mile-an-hour wind that's pulling the shingles out of your hand, or just putting the boards up because there's a gale that has come up all of a sudden, or whatever it is, pulling your neighbors out of the sand. You can't bring your own stuff and try to control your environment out there. You have to let it go. And I think that's part of the neighbors deal, what makes them so essential to this whole way of living out there.

Basic Elements of Dune Shack Society

This interview with Emily Beebe about her personal experiences clarified for me the basis for dune shack society, the "parameters" as she states it. Three basic elements, in combination, delimit what I will call "dune shack society": the dunes, the shacks, and long-term shack residents. There are other elements as well, such as ways of living, belief systems, sociopolitical struggle, and so forth, as discussed elsewhere in the report. But these three are basic parameters. First are "the dunes" themselves. The Backshore system of dunes with its sands, storms, vistas, and other natural features is viewed as a distinct and unique place for Beebe and other shack residents. This place of raw, uncontrolled nature is what initially drew her, a place to discover and experience, and is what she now must adapt to as a dune dweller. Second are "the shacks" themselves, cultural places erected on the dunes, scattered and solitary, rustic and weather-beaten, relatively low-impact shelters allowing people to be long-term dwellers on the dune system. Beebe noticed and wondered about them, imagining them as ocean sentinels, survivors, absorbers of special energy, silently holding a wealth of personal stories. The structures drew her also, like the dunes. Third are the "long-term shack residents" and their respective coteries of shack users. Beebe and Simon discovered this social group by gaining entrance into one of the shacks, by receiving the key to long-term residency by the Seashore. Conversely, it may be equally said that dune shack society discovered Beebe and Simon. Through this mutual encounter, the two newcomers were incorporated into an existing social group that is centered in long-term shack residencies. These three are distinct elements – the dunes, the shacks, and the long-term shack residents. Yet when connected over time, they have become the basis for the emergence of a unique social group. It was a nonstandard, sometimes hard-to-see community of self-avowed nonconformists, sharing a history, traditions, cultural patterns, and social identity, on a backshore edge of 21st century America.

Counting Shack Populations

Emily Beebe's story recounted her entry into dune shack society. But what was this society of dune dwellers? Who were the dune dwellers? How many were there? How were they connected? To answer these questions, I began interviewing the heads of the dune shacks about the people who used their shacks.

Early on I learned that the Seashore had no tallies of shack residents or users. As I became immersed in the project, I realized that no one maintained a list of dune shack residents or users. There were no censuses, no registers, and no rolls, such as voter rolls of local election districts or tribal membership rolls of Native American governments. Dune shack society was not organized to compile such lists. It had no internal government or fully representative bodies. Two nonprofit organizations tallied counts of users for four shacks under their care. Provincetown Community Compact had user estimates for C-Scape, the one shack it managed. Jay Critchley and Tom Boland of the Compact graciously shared them. Peaked Hill Trust had estimates of its membership and some tallies of users of the three shacks it oversaw. Otherwise, no entity, agency, or person kept numeric tabs on dune shack society. No one had ever documented its size or composition.

The federal government did maintain a list of shack occupants with whom the Park had legal relationships. The list was an electronic spreadsheet entitled, "Occupant and tract and expiration list, 2004," and it listed "people and organizations with whom the NPS has a legal relationship for occupancy and or use of the dune shacks… with the exception of the one private landowner." It named eighteen shacks, twenty-eight persons, and three organizations. A footnote advised, "This list is not for general distribution as it includes Privacy Act protected information." The spreadsheet identified people and organizations with whom the Park had one of four types of legal relationships as of March, 2004: a reservation of use (9 shacks, 15 people), a lease (3 shacks, 6 people), a cooperative agreement (2 shacks, 3 people), a special use permit (3 shacks, 3 people), or private ownership (1 shack, 1 person). Essentially, the list identified legally-responsible occupants for each shack. The 28 occupants were actually 26 persons, as one person's name was listed for three different shacks. The count (18 shacks, 26 people) represented a conservative enumeration of dune shack society. I used the list as a starting point for interviews.

During interviews with shack heads, I attempted to more fully document the numbers of shack users and their relationships with one another. I observed that there seemed to be a core of long-term residents associated with each shack. This core might be counted with some precision because it had a stable structure, usually kinship based. I also observed that there was a larger set of short-term users, the coteries of this core, comprised of friends, guests, drop-ins, or awarded renters. This set of users appeared to be more fluid, their names and numbers changing from year to year depending upon the circumstances of the long-term residents, such as their health, employment, life stages, and activities, among other factors. This was the general structure of the population – a fairly stable core of long-term residents with a larger, more fluid set of short-term users who were guests of the core residents.

Interviews with shack heads provided me a way to document the size and composition of these types of users. During interviews I asked shack heads, "Who generally uses this shack, and how are they connected to the main shack residents?" This general question provided information about the core network of users of shacks. I also asked, "How many people use the shack during a year, such as the last twelve-month period?" Most shack heads easily remembered what had happened in their shacks during the last twelve months. This general question provided information for documenting users other than people in the core. Accordingly, I compiled this information to form a conservative estimate of the dune shack residents and users. The population of dune shack users, broken out by shack and type of user, appears in the following table.

Dune Shack Single-Year Population *

Shack Name and Type	Shack Heads	Core Network	Other	Estimated Totals
Shacks with Family Caretakers				
2. Beebe-Simon Shack	2	34	82	118
3. Adams Guest Cottage	2	18	10	30
4. Adams Shack		*(Included with No. 3)*		
5. Champlin Shack (Mission Bell)	2	8	10	20
6. Malicoat-Lord Shack	2	10	20	32
9. Tasha Shack	7	31	50	88
10. Jackson Shack	2	9	44	55
11. Fowler Shack		*(Included with No. 10)*		
12. Clemons-Benson Shack	2	17	54	73
13. Schnell-Del Deo Shack	3	6	20	29
15. Schuster Shack	1	1	15	17
16. Isaacson-Schecter Shack	2	15	60	77
17. Wells Shack	1			1
18. Dunn Shack	2	4	6	12
19. Armstrong Shack	2	20	15	37
Shacks with Nonprofit Caretakers				
1. C-Scape	2	10	42	54
7. Werner Shack (Euphoria)	12	20	44	76
8. Gelb-Margo-Zimiles Shack	*(Included with No. 7)*		44	44
14. Werner Shack (Thalassa)	*(Included with No. 7)*		44	44
All Shacks and Types	44	203	560	807

* *Residents and other users during the previous twelve months. Over time, the set of persons using shacks is greater than the users during a single year.*

Based on interview materials, my rough estimate of persons using the dune shacks during the most recent year was about 807 people. Of these, I placed 44 users in a category of "shack heads," that is, residents with the primary responsibility for overseeing a shack, typically the heads of a family or of a nonprofit group. I placed 203 users in a "core network" of long-term residents, that is, consistent users of a shack, typically members of an extended family associated with a shack, or regular active members in a nonprofit group with caretaker responsibilities. I placed 560 users in the "other" category, representing short-term users, typically friends invited to the shack, persons awarded "shack time" by a nonprofit group, or more occasionally, drop-in strangers. In total, the estimate of 807 users during a year's time is about thirty-one times the starting list of twenty-six legal occupants.

The actual set of persons directly connected to dune shacks through residency and use is larger than 807 people. How much larger is difficult for me to say with accuracy with the information I have collected. The core set of users (about 247 people in the above table) changes less from year-to-year than the "other" category of short-term users (about 560 people last year). If one assumes that, during recent years, the pool of "other" short-term users was two or three times greater than the set of last-year's short-term users drawn from that pool (that is, assuming the set of short-term users was between 1,120 and 1,680 people), and if one assumes a core of 247 users, then the people directly associated with the dune shacks through residency and use during recent years numbered somewhere between 1,367 to 1,927 persons. More information would be needed to assess the reasonableness of these assumptions. I would not be surprised if the pool of long-term and short-term users over the past several years might be found to be larger. However, my best estimate is that dune shack society currently numbered between about 1,367 to 1,927 people. Given its shaky basis, I anticipate this estimated size of dune shack residents and users to generate lively debate. But it's a starting point toward understanding the social dimensions of dune shack society.

Family Networks: The Primary Organization of Dune Shack Society

Interviews with shack residents quickly established the social core of dune shack society – family networks. I observed a common feature among shacks not managed by nonprofit organizations: shacks typically housed members of families, that is, people connected through kinship. The most common "customary rule" for qualifying as a core user of a dune shack was being closely connected by descent or marriage to the shack heads, figured through both maternal and paternal sides (a bilateral system). In filling their shacks with people, the heads of dune shacks commonly opened their doors to members of their extended families. Over time, it was common for several generations of an extended family to use a shack.

During interviews, several shack residents concurred that families lay at the core of shack society. There were certainly more kinds of shack users than simply family members. But families were central. With the affirmation of these observations, I added a component to interviews, the drawing of kinship diagrams. The diagrams proved an efficient tool for identifying many regular shack users and for establishing their relationships with the shack heads and themselves. These kinship networks are presented below for each family-based shack where the information was collected, with a short descriptive narrative. The symbols in the diagrams are as follows:

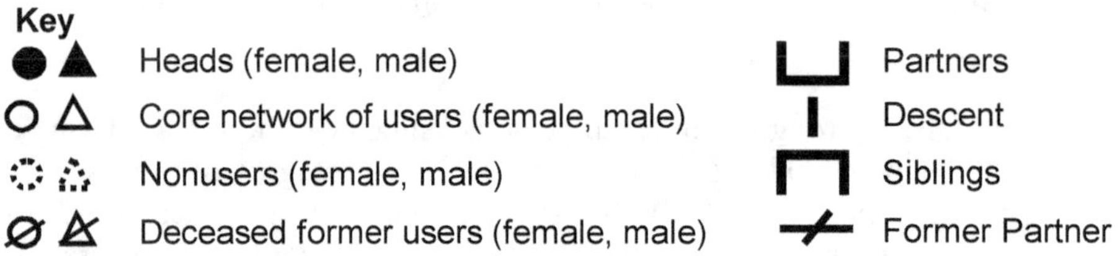

Key

● ▲ Heads (female, male)　　　　　　　　⊔ Partners

○ △ Core network of users (female, male)　　| Descent

⊙ ⊿ Nonusers (female, male)　　　　　　　⊓ Siblings

⊘ ⊿̸ Deceased former users (female, male)　　⫽ Former Partner

1. The Adams Shacks: Heads and Core Network

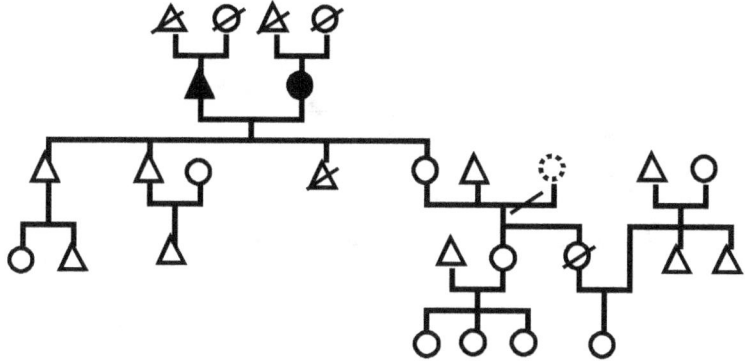

The current heads of the two Adams shacks were David and Marcia Adams (depicted in black, above), whose second home was in Kalamazoo, Michigan. Five generations of people from this extended family have used the two shacks. These include David and Marcia, their four children, two of their children's partners, six grandchildren, and four great-grandchildren, as well as David and Marcia's four parents. Included in this core network of users was an additional household of four linked through an adopted great-granddaughter.

2. The Champlin Shack: Heads and Core Network

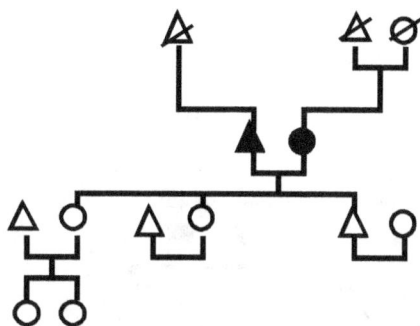

The current heads of the Champlin shack were Nathaniel (Nat) and Mildred Champlin, whose second home was in Bristol, Rhode Island. Four generations of people from this extended family have used the shack. Current users include Nat and Mildred, their three children, their children's three partners, and two grandchildren. In addition, Nat's father and Mildred's parents used the shack.

3. The Malicoat-Lord Shack: Heads and Core Network

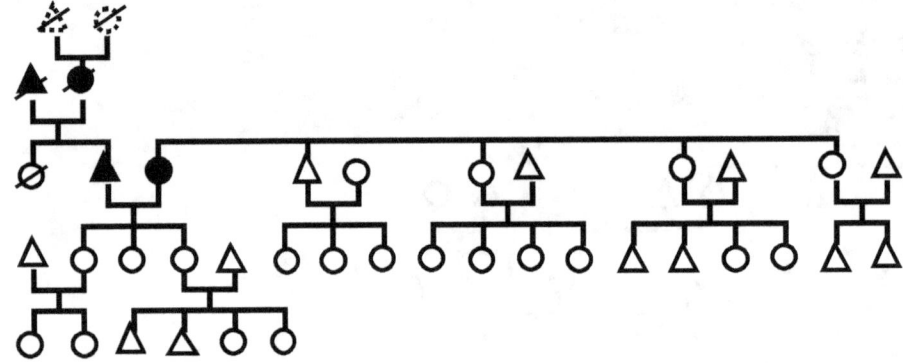

The former heads of the Malicoat-Lord shack were Philip Malicoat and Barbara Brown Malicoat of Provincetown. Currently, the heads were Conrad Malicoat (their son) and his partner, Anne Lord. Five generations of this Provincetown family have used the shack. The most regular users have included Conrad and Anne, their three children, two sons-in-laws, and six grandchildren. In addition, Anne's four siblings with their partners, and thirteen nieces or nephews used the shack. Conrad's sister (Martha M. Dunigan) also used the shack before she passed on. Conrad's great grandparents listed above were Harold Haven and Florence Bradshaw Brown, both former directors of the Provincetown Art Association (1928-32 and 1932-36 respectively).

4. The Tasha Shack: Heads and Core Network

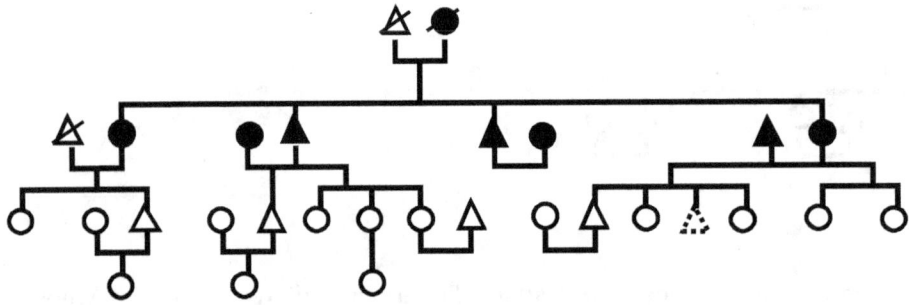

Formerly, the head of the Tasha shack was Rose Savage Tasha (Sunny), a Provincetown resident who used the shack with her husband (Herman Tasha) and her four children, Carla, Carl, Paula, and Paul. With her passing, the shack remained in the family with its use decided jointly among the four children and their partners, depicted above as a consortium of seven. There have been four generations from this extended family using the shack, including Sunny and Herman Tasha, their four children, their children's partners, eleven grandchildren with four partners, and three great-grandchildren. All resided on Cape Cod, primarily in Provincetown, Wellfleet, and Truro.

5. The Schnell-Del Deo Shack: Heads and Core Network

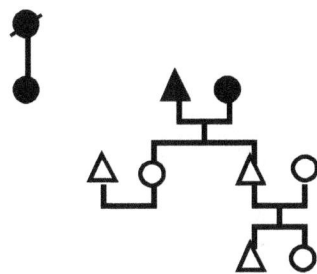

The former head of the Schnell-Del Deo shack was Jeanne Chanel (Frenchie). Currently, the principal head was her daughter, Adrienne Schnell (Schatzi). Two co-heads were Salvatore and Josephine Del Deo, close friends of Frenchie and Schatzi, who were bequeathed the shack by Frenchie. Provincetown has been home for all four. From this network, four generations have used the shack, including Salvatore and Josephine's two children, their children's partners, and two grandchildren.

6. The Armstrong Shack: Heads and Core Network

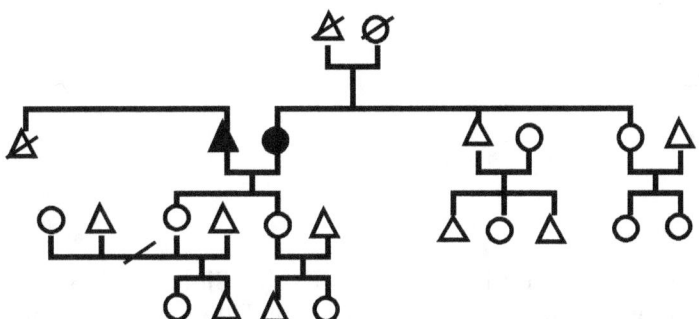

The current heads of the Armstrong shack were David Armstrong and Connie Eshman Armstrong, whose second home was in Maine. Four generations of this extended family have used the shack. Regular users have included David and Connie, their two children, three sons-in-law, and four grandchildren. Additional users have included Connie's parents, Connie's two siblings with their partners, and five nieces or nephews. David's brother (John) was a regular user before he passed away last year. The extended family represented fourteen households currently using the shack.

7. The Clemons-Benson Shack: Heads and Core Network

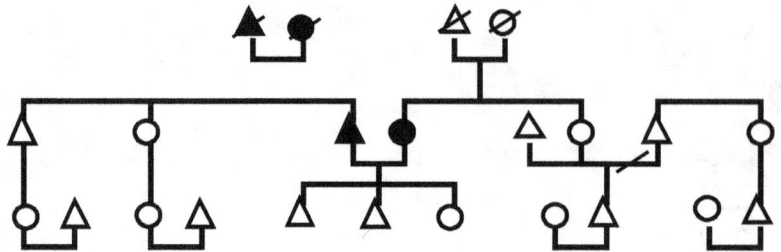

The current heads of the Clemons-Benson shack were Peter Clemons and Marianne Benson, whose second home was in the Boston area. The former heads were Andy Fuller and Grace Bessay, who were like grandparents to the Clemons-Benson children, with Andy being the godfather of the second son, his namesake (David Andrew Clemons). Three generations from this extended family have used the shack, including Marianne's parents, Peter and Marianne, and their three unmarried children. Others in this core of users included Peter's two siblings and two nieces with their partners, and Marianne's sister, her sister's two partners, and a nephew with a partner. Other users connected by kinship included a brother-in-law's sister, her son, and her daughter-in-law.

8. The Beebe-Simon Shack: Heads and Core Network

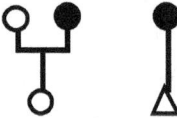

The current heads of the Beebe-Simon shack were Emily Beebe and Evelyn Simon, jointly chosen by the Park as residents for the Fleurant shack in 1993. Also included in the core network of consistent users of the Beebe-Simon shack were two children and Emily's partner (pictured above), as well as thirty-one people not shown in the diagram – a mother, two siblings, and twelve close friends of Emily, and ten relatives and six close friends of Evelyn.

9. The Gelb-Margo-Zimiles Shack: Heads and Core Network

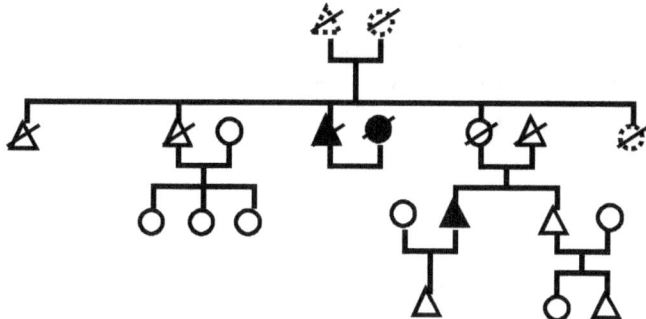

The former heads of the Gelb-Margo-Zimiles shack were Jan Gelb and her partner, Boris Margo, and their nephew, Murray Zimiles, reared like a son and bequeathed the shack by Boris. Three generations from this family have used the shack, including seven siblings or sibling-in-laws in Jan and Boris' generation, seven in Murray Zimiles' generation, and three in the third generation. Members of Murray Zimiles' extended family have continued to use the shack annually, among others with allotted time (see Chapter 8). Currently, Peaked Hill Trust served as caretaker of the shack under an agreement with the Seashore.

10. The Dunn Shack: Heads and Core Network

The current heads of the Dunn shack were John Dunn (Scott) and Marsha Dunn, whose second and third homes were in New Mexico and Wellfleet. The Dunns were awarded a lease on their shack by the Park Service in 1993. There were two generations from this family who have used the shack, including Scott and Marsha, their son, and their son's partner.

11. The Schuster Shack: Head and Core Network

The current head of the Schuster shack was Lawrence Schuster. Lawrence lived in the shack year-round. Of his family, the shack's other common user was his daughter, who lived in Provincetown.

12. The Isaacson-Schecter Shack. (No diagram collected.)

The current heads of the Isaacson-Schecter shack were Gary Isaacson and Laurie Schecter, whose second home was in Florida. They were awarded a lease on their shack in 1993.

13. The Wells Shack. (No diagram collected.)

The current head of the Wells shack was Ray Martin Wells. I did not conduct a formal interview with Ray Wells to document the current users of her shack.

14. The Jackson Shack: Heads and Core Network

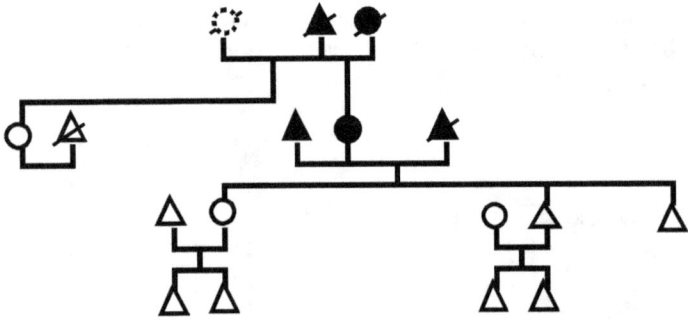

The current heads of the Jackson shack were Zara and Samuel Jackson, whose second home was in the New York area. Formerly, the shack was headed by Zara's parents, Alice Amitina Malkin and Martin Malkin of the New York-New Jersey area. When they died, Zara became head, eventually with Irving Ofsevit, her husband. There have been four generations of people from this extended family using the shack. Users have included the five people previously listed, as well as Zara's three children, two of the children's partners, and four grandchildren. Another relative connected to the shack has been Ray Wells, Zara's half-sister, who at times managed the shack with her husband, Nick Wells, at Zara's request. Ray and Nick Wells owned a neighboring shack.

15. The Fitts-Walker Family Network

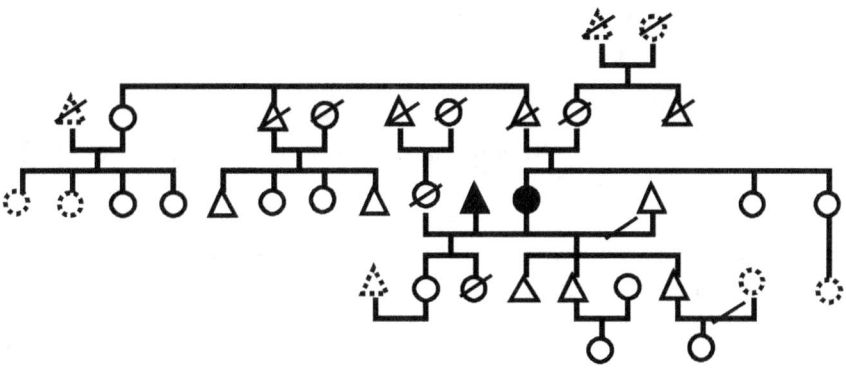

The extended family of Bill Fitts and Harriet Walker Fitts (Hatty) is illustrated above. In 2004, Hatty Fitts was the legally-responsible occupant for three shacks (two Werner shacks and the Margo-Gelb-Zimiles shack) for Peaked Hill Trust and OCARC, through agreements with the Park Service. Hatty and Bill Fitts were part of an extended family centered in Provincetown whose members have used dune shacks as friends and guests of shack residents over the years, although none has owned shacks. In the Walker line, four generations of Provincetown residents have used the dunes and shacks. The diagram shows eighteen people currently living who have used shacks, and nine other users who have since passed away. The living users include Bill and Hatty, their four children, and two grandchildren, as well as Hatty's two sisters, aunt, and six cousins. In 2004, Bill and Hatty lived in Provincetown, and three of the four children were local (Truro or Provincetown). Bill and Hatty Fitts have used several shacks, including the Gelb-Margo-Zimiles, Dunn, and Malicoat shacks, among others. During the 1980s, Bill and Hatty played central roles in the administrative effort to find the shacks eligible for listing as historic places. They were prime movers in the reconstruction of the Jackson shack after it burned in 1990.

Networks of Friends: The Secondary Organization of Dune Shack Society

While family networks comprised its core, dune shack society had another essential segment – networks of friends who used shacks as guests of core families. As I visited shacks, I commonly encountered shack users other than kin, people who were introduced as friends of the family. At times in the life history of certain shacks, the majority of shack users were friends of the core family, and not kin. I will present two examples to illustrate the networks of friends that used shacks: (1) friends drawn from the extended family of Harriet Walker Fitts over the last 84 years, and (2) friends connected to the Clemons-Benson shack during the last twelve months. There were many other friendship networks in addition to these two.

The family of Harriet Walker Fitts (Hatty), described above, provides a case illustrating the interconnections of some old Provincetown families with the dune shacks through friendships with core shack families. In the Fitts line, these connections extended back at least 84 years, to 1917. Hatty Fitts said she herself has used shacks since she was four years old up to the present (she was 64 years old), although no one in her family had ever owned a dune shack. She described the history of her family's connections with the dune shacks through friendships:

My grandmother and grandfather Gaul started coming here in 1917. My grandfather was a musician. My grandmother was a writer. They were all part of the Eugene O'Neill group, the playhouse group. That's really when they first started [using the dunes]. They used to be part of the group that would go out when Eugene O'Neill went out there. They would trek out and visit with him.

My grandparents never really stayed out there for any long periods of time. They were just visitors. My mother was Ione Gaul Walker. When she was growing up, she also was involved as a visitor. She had friends who lived out there. She became friends with a number of coastguardsmen who were at the Peaked Hill Station. Some of them lived in the East End of town, which was where she always lived. She knew them as local neighbors in town and would go out to visit with them in the dunes when they were on duty.

I started coming [to the dunes] in 1943, and have come ever since, and again, had friends. At that point there were two places we would go on a regular basis, the Jones shack, which is now leased by the Dunns, and the Malicoat's shack. We used to go out to visit. At the Jones shack we used to actually stay there, not my whole family, but my sisters and I would stay there with other friends. The Joneses were very good friends with the Thomases. Jimmy and Edith Thomas lived down the street from us. Jonathan was the son. Fred was the grandfather. Jimmy had a model A Ford truck jerry rigged for the dunes, a dune buggy. That's how I learned to drive, when I was thirteen. In the late 1940s, my uncle Philip Walker started going out on a regular basis. We were also friends with Art Costa who started the dune tours. They were all great buddies together. At that time you could drive anywhere on the dunes.

We used to have regular get-togethers with the kids from the East End. We would make blanket rolls and pack out to the dunes and sleep out there. We had food. We never had campfires in the dunes. We always took sandwiches, nothing that required any cooking. We'd do this on a pretty regular basis because living in the East End we were right down from Snail Road. At that time there was no main highway. Snail Road was a wooded path that went from what is now Highway 6A all the way into the dunes. You could drive it. That's the way people got into the dunes. Now there's just no way you could get up that hill. But then, there was no hill like that. You could get up without any problem at all. It was quite a trek through the woods, where we went blueberrying and so forth. My grandmother used to go out with us also and spend the night in her bedroll. If we got permission to go to the Jones', that's where we would go. Otherwise we would just go up on the big dune.

Now I have children, and Bill's daughter was virtually born using shacks. My kids have been going out and visiting people and staying in shacks almost since they were born, maybe one or two years old. And my granddaughter has now stayed out in the dunes a couple of times. So we're working on five generations.

As far as the shacks, the only ones I really got to know [as a child] were the Jones shack and the Malicoat shack. I was aware of Euphoria but I really didn't get to know that one until I was an adult. Hazel [Werner] was also a friend of my parents and had invited them to use the shacks too.

That's one of the things about Provincetown, this woven pattern of many, many families over generations. You try to take one family in town and keep it separate – it doesn't work. Everybody has connections to everybody else. And that's how people got out on the dunes in the first place. It was because of these connections. They are very strong connections, very tight friendships that have gone on generation to generation. Then those friendships lead to marriages that intertwine everybody.

As shown above, the Gaul-Walker-Fitts family traditions directly descend from the activities of historically-significant figures on the dunes, including Eugene O'Neill, Hazel Hawthorne Werner, Phil Malicoat, the coastguards at the Peaked Hill station, among others. Even though none ever personally owned a dune shack, members of this extended family used the shacks through friendships with shack owners. Families like hers have continued to be important parts of dune shack society, using the shacks alongside core residents, and in more recent decades, working to preserve the shacks through political action and nonprofit organizations.

The second case illustrating friends using shacks is the Clemons-Benson shack, and its neighbor, the Fowler shack. One August evening, I encountered seven people at these shacks, four in the Benson-Clemons family and three friends of the family. The Clemons-Benson shack and Fowler shack are neighboring cottages, sitting several hundred steps from one another in the central cluster of dune shacks. They are reached by walking straight out Snail Road and taking the left hand branch almost to its end, the route called by some the O'Neill Path, the historic way to Eugene O'Neill's old summer home. The current head of the Fowler shack is Laura Fowler, an elderly woman living in Florida who no longer can visit her shack. She was a long, close friend of people in the neighboring shack: Grace Bessay, Andy Fuller, Peter Clemons, and Marianne Benson, and before them, Doey and Al Fearing. For several years, Laura Fowler has asked Peter Clemons to care for her shack, maintaining and using it. My interview in August began in the living room of the Fowler shack and then moved over to the neighboring Clemons-Benson shack. This mirrored the current use pattern of the paired shacks, as the Clemons-Benson family and invited guests were using both, shifting back and forth between them depending upon who was staying any particular week. In 2004, the Clemons-Benson shack was being used as a painting studio as well as for lodging.

Four of five members of the Clemons-Benson's nuclear family were present that August evening – Peter Clemons and Marianne Benson and two sons, David Andrew Clemons (Andy) and Thomas John Clemons, single young men who had grown up at the shacks during summer. In addition, seated next to Benson and fully participating in our lively interview was David Forest Thompson, a tall, robust artist who was a friend and work associate of Peter. I learned that Thompson and Peter were painting together that summer. Thompson was working on a project to publish images of every dune shack on the Backshore, past and present. Some of his shack images were on display at the Backshore Gallery, Peter's art gallery at the center of Provincetown's art district. Thompson hosted a showing of some of Peter's artwork at his combined gallery-salon in Boston. Partway through our interview, a tall, muscular man in a tiny swimsuit entered the shack, toweling off from a late afternoon swim. This was Ray Carpenter, a friend of Thompson. He hung on the sidelines, an interested listener. Then a seventh person showed up, hiking across the dunes, young Harry Upsahl, a friend of Thomas and Andy. With a smile, he stated he was a ten-year Provincetown resident, but still considered a newcomer (a "washashore") by some. After the interview, the seven ate dinner together and conversed, and then lined up, shoulder to shoulder in front of the shack, posing for a group photograph for my project (see Fig. 9).

My encounter at this dune shack was not unusual. The group of people at the shack was a mix of family and friends. This was a common use pattern for shacks. Lawrence Schuster, a year-round dune shack resident and cultural anthropologist by training, summed up the traditional use pattern for shacks: "People had shacks. They stayed there. And they had friends out. That's the tradition." Schuster said there had never been a traditional use of artists-by-the-week, or of a lottery system for an organization's members. This was not the traditional way that shacks worked. The short-term, programmed experiences of artists and other visitors represented a different arrangement, products of cooperative agreements between nonprofit organizations and the federal government. Such programs were introduced and not part of the shacks' usual cultural use patterns, he said. Traditionally, the people who had the shacks "stayed there" and "had friends out."

I wanted to understand the set of friends who stayed at the Clemons-Benson and Fowler shacks and how they were connected to the heads who "had the shacks," using Schuster's language. Because she was willing, I requested help from Marianne Benson. I requested her to make a list of people who had used the Clemons-Benson and Fowler shacks during the last twelve months, to the best of her memory. She called me a week later with her list. I was astounded, perhaps because of my expectation of solitude at shacks. There were over fifty names, not counting relatives. "Who are all these people?" I asked. She laughed and simply said, "friends." Working together, she and I traced out the linkages between the people on her list and her nuclear family of five members. The task took several hours, as there was a short story to tell about each linkage, its history and nature. Upon conclusion we had charted out a network of friends, illustrated below, who had used the Clemons-Benson and Fowler shacks last year. More details about links with the Clemons-Benson family are summarized in the accompanying table. In this table, Fr means "friend," Hu means "husband," Wi means "wife," Si means "sister," Br means "brother," Da means "daughter," and So means "Son."

Network of Friends Using the Clemons-Benson and Fowler Shacks Last Year

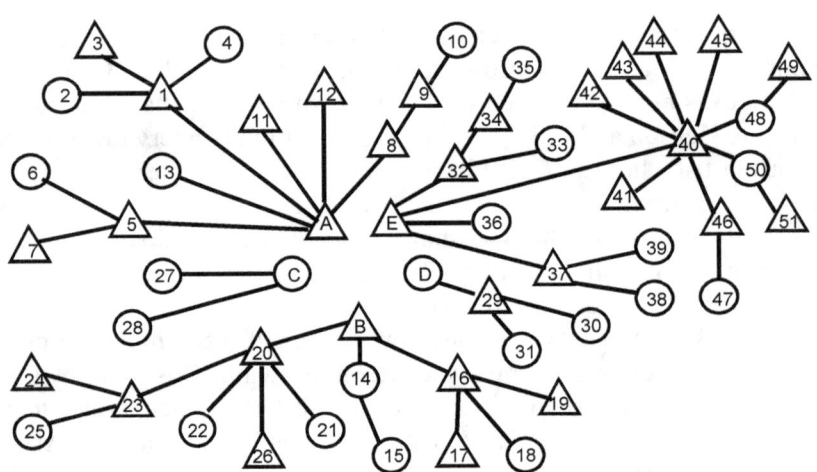

A. Thomas Clemons				
1. Fr (Harry)	2. FrMo (Olga)	3. FrFa (Peter)	4. FrSi (Oden)	School
5. Fr (Adam)	6. FrMo (Ginny)	7. FrFa (Jim)		School
8. Fr (Kyle)	9. FrFr (Ben)	10. FrFrSi (Maggie)		School
11. Fr (Trevor)				School
12. Fr (James)				School
13. Fr (Annie)				School
B. David Andrew Clemons				
14. Fr (Julia)	15. FrMo (Anka)			Provincetown
16. Fr (Perry)	17. FrBr (Cyrus)	18. FrMo (Nora)	19. FrFa (David)	School
20. Fr (Reif)	21. FrFr (Kyle)	22. FrMo (Judith)	23. FrFa (Peila)	School
24. FrFaBr (Jasper)	25. FrFaFr (Victoria)	26. FrFr (Chris)		
C. Elizabeth Clemons				
27. Fr (Lizzie)				School
28. Fr (Sue)				School
D. Marianne Benson				
29. Fr (Ben)	30. FrMo (Lisa)	31. FrMo (Deb)		School
E. Peter Clemons				
32. Fr (Jerry)	33. FrWi (Phylis)	34. FrSo (Carl)	35. FrSoWi (JoAnn)	Work
36. Fr (Rebecca)				Work
37. Fr (Nick)	38. FrWi (Ann)	39. FrDa (JoJo)		Church
40. Fr (David Thompson)				Work
40. David Thompson				
41. Fr (Sonny)				
42. Fr (Ray)				
43. Fr (Richard)				
44. Fr (Bill)				
45. Fr (Frank)				
46. Fr (Paul)	47. FrWi (Debbie)			
48. Fr (Marsha)	49. FrHu (Skip)			
50. Fr (Brenda)	51. FrHu (Kevin)			

* Source: Marianne Benson, August 2004

The diagram shows the five members of the Clemons-Benson at the center: the household heads, Peter (E) and Marianne (D), and their three young-adult children, Thomas (A), David Andrew (B), and Elizabeth (C). Friends that used the shacks during the last twelve months are shown. There were fifty-one, although Marianne stated she might have inadvertently forgotten some. Obviously, they did not come all at once, but singly or in small groups distributed over the course of the year. They were linked to the core family through school, work, and church. The figure and table identify the single, most-direct linkage between people and the core. For example, six people who came to the shack were school chums of Thomas, and they brought with them seven others, relatives-of-friends or friends-of-friends as reckoned from Thomas. "School" also provided the links for friends brought by the other two children. Marianne stated that the parents of several of her children's school friends also were friends of Peter and herself. However, these additional connections are not depicted in the diagram, which intends to show the initial link with the family. "Work" was the most frequent link for six friends of Peter. "Church" was a connection for three friends. One work-related friend, David Thompson, brought with him eleven additional users.

The Clemons-Benson and Fowler shacks received a relatively large number of visitors compared with other shacks last year, in my assessment of information provided by other shack residents. Because the Clemons-Benson resided in two fairly commodious shacks, there was room for accommodating visitors. Lawrence Schuster, with one bed in his shack, said he entertained a few guests but had little room for overnight visitors except in sleeping bags outside on the sand, which happened on occasion. So the size of a shack was partly related to numbers of friends who used it, though not determinant. The Tasha shack, the tiniest of all the shacks, drew relatively large numbers of visitors over the course of a summer, probably due to the connectedness of the Tasha family in Provincetown and the "open door" policy of the Tasha family, discussed elsewhere.

The Clemons-Benson shacks currently drew many guests from a pool of schoolmates. This was related to the older ages of the three children and the close friendships developed at their boarding schools over time. The frequency of such visitors could change as their children age, establish families, and perhaps move to other places. These types of life history dynamics in shack resident composition are discussed in Chapter 4.

In addition to large shack size and mid-life family stage, the Clemons-Benson family was gregarious, welcoming of visitors and spontaneous gatherings of people brought in by children and friends. However, I also heard Peter say a couple of times that he had to schedule time alone at his shack in order to paint. So there was a mix of solitude and social gatherings. Some shack residents were not as gregarious as the Clemons-Benson. Getting away from people was a central use of some shacks, such as Ray Wells' shack. "Nobody bothers me at my shack," she told me. Historically, Charlie Schmid's shack was used primarily as a single-person home for him, where he observed swallows and other natural cycles on the Backshore. To many, he seemed to be a hermit. Fewer friends would be likely to be in evidence at these types of shacks.

Drop-In Strangers: The Occasional Visitors

Drop-in strangers comprised another set of users of the Backshore dune shacks, and accordingly, might be considered another segment of dune shack society. I have no firm estimates for the numbers of users in this category. However, this type of use is without a doubt substantially less than use by extended families and the friends of extended families. Drop-ins may actually be infrequent for most shacks. That it occurs is shown above by the narrative of Emily Beebe earlier in this chapter, who recounted walking among the shacks, not wanting to encroach on their privacy, but occasionally sitting on their decks. Most occasional dune users like hikers, fishers, and tour groups probably leave the shacks alone because they perceive the shacks to be private residences. They are not perceived to be public facilities like picnic tables, open for general use unless otherwise restricted. A couple of shacks, I noted, displayed signs saying, "private property" and "no trespassing," but most did not. I was told that Park Service employees were instructed not to enter a dune shack without the permission of residents. Bill Burke, the Seashore's historian with a keen interest in the shacks, told me he had not been inside many of the shacks yet. He was not a drop-in stranger. There is little historic evidence suggesting that drop-in use by strangers has ever been a substantial category of dune shack use, notwithstanding the shacks' asserted origins in the "hospitality huts" along Cape Cod's Outer Beach (see Chapter 2). Based on shack residents' reports, drop-in use by strangers has never been a large use of shacks. More use by drop-in strangers probably occurred during the turbulent sixties and seventies, as discussed in later chapters. Currently, use by drop-in strangers is probably low for most shacks.

Most shack residents I interviewed stated they were uneasy of strangers who used shacks without permission, primarily because of the potential for accidents, vandalism, theft, and fire. Customarily, shack residents looked out for one another's shacks, discussed in Chapter 5. Shack residents typically locked their shacks when gone for extended periods to protect the premises from vandals, unless they were expecting guests. Two exceptions were the Tasha and Isaacson-Schecter shacks with "open-door" policies. But these were exceptions to the customary practice of protecting shacks from misfortune due to the carelessness or mischief of strangers. Nevertheless, it appeared to me that shack residents displayed a high tolerance for intermittent use of shacks by drop-in strangers, as long as nothing was broken and courtesy shown. It happened. Strangers sat on decks. They pumped water from shack wells. They might even enter an unlocked door to look around. Shack residents expected this on an occasional basis. They were among the many expected, uncontrolled events inherent in having a life on the dunes, far from town and away from the public eye.

Friendly Renters and Barters: Established Social Forms

I concur with the general observation of Lawrence Schuster, that people had shacks, stayed there, and had friends over. That was the principal traditional use. However, historic patterns of shack use were more complicated than that, as attested by shack residents I interviewed, including Schuster. At times, some shacks rented. Historically, renters comprised another segment of dune shack society. Some dune shacks rented on an occasional basis when not in use by the core users. This was an old tradition on the Backshore, though not the predominate one. Historically, the rental arrangements for dune shacks commonly were relatively informal, greatly subject to personal relations between the shack owner and the renter, and not tied to the market of summer cottage rentals elsewhere on Cape Cod. Agreements commonly were extensions of personal relationships between the shack head and the renter, rather than a formal contract between strangers. However, for a time the Werner shacks and some of the early coastguard shacks did rent to strangers on a more formal basis. Currently, shack residents said that Seashore rules prohibited or substantially limited renting shacks, except for the shacks cared for by nonprofit organizations. Therefore, renting was a use pattern of family-centered shacks that primarily existed prior to the Seashore.

Rental arrangements were possible when dune shacks were used seasonally or intermittently by core residents. Families occupied them during some seasons and not in others. Families spent more time in the shacks on some years than on others, depending on circumstances. For particular shacks and families, there arose times of extended vacancies by principal users, when the shack might lay fallow. In these instances, some shack residents made arrangements with people to care, maintain, and use the shack in their absence. There were several types of arrangements – labor exchanged for rent, other exchanges of services, and outright gifts. But not uncommonly, a nominal monetary rent was charged for the temporary residents. Because these were often based on personal relationships, sometimes the rents were not even collected. For example, Barbara Baker lived in two of the dune shacks when she had young children. She said she stayed in the Jackson shack for a couple of years and in the Wells shack for one year. She was supposed to be renting from the owners, however she believes she wasn't ever charged because she and her husband had so little money. However, they maintained the shacks for the owners while there. Examples of shacks commonly rented included the Werner shacks, the Jackson shack, the Wells shack, and the Braaten shack (Schuster shack). Hazel Werner came to rent her shacks when she fell into poor heath. Lawrence Schuster reported he was a renter of a Werner shack and later, the Braaten shack when it lay empty.

As illustrated by the Baker case, many shack users were given time in a shack in compensation for work on the shack. This was a form of barter, an exchange of shack occupancy for in-kind services. This historic use of shacks may be even more common than renting for money. The exchanges were examples of transactions within the traditional, informal economy of cash-poor Provincetown, called an "underground economy" by Jay Critchley, a shack user. Labor exchanges for shack time has been an extremely important mechanism for the survival of the dune shacks, as discussed in greater detail in Chapter 11. As shown in Beebe's account of her initiation into dune shack society, people volunteered to help her renovate Leo's shack. She rewarded many with opportunities to stay at the shack. This was one of many examples of this type of arrangement I heard of during the course of interviews. When a person voluntarily worked on a shack, that person often was invited to use the shack in the future. Some workers specifically asked for shack use in lieu of monetary payment. The reciprocity of shack use for labor most often occurred when friends and associates volunteered work and then received chances to use a shack from the shack's heads. Of course, family members commonly worked on their own shacks while staying in them. However, this labor was rationalized as part of familial obligations, rather than a type of reciprocal barter of labor-for-shack use.

Peaked Hill Trust and Provincetown Community Compact, two nonprofits associated with the shacks, commonly used barter in their programs for caring for and managing shacks. Members or associates of the nonprofits who voluntarily worked on shack upkeep and maintenance commonly were compensated with shack occupancies when shacks were not used by winners of lotteries and artist-in-residence grants. The arrangement for the C-Scape shack program was described for me by Tom Boland, who managed the shack used by the Provincetown Community Compact:

I have a running list that I keep of people who have helped out. If someone cancels or leaves early or comes late, it's pretty common, I'll call them and say, "I have a night here and I have a night there," and if their schedule allows they'll go out. Also, if people do a major job, they get time. Like I have two women going in October. One of them is a contractor. Last spring, not this past spring but the one before, she and I went out there with another friend of mine who works here. The three of us built a deck, in exchange for her to get a week.

This arrangement was more formalized with shacks cared for by the Peaked Hill Trust. As described below, about a dozen caretakers in the organization maintained the shacks and transported shack users in exchange for shack time. This was in lieu of workers paid with money. The nonprofits relied on this type of traditional arrangement with workers for maintaining shacks.

Some shack owners, such as the Champlins and Murray Zimiles, said they avoided renting their shacks. Zimiles stated that the shack was like "holy ground," and he did not believe in renting it. The Champlins and Adams said they did not trust people other than family to care for the shacks, because of risks like fire:

Mildred Champlin: The Adams were gone for several years because David [Adams] was working on his doctorate, and Marcia was teaching, so they didn't come out here. We took care of their house. We had our friends use that house and we maintained it. We did rent it out to Conrad Malicoat's family who's next door. This was, oh, in the very early 60s.
Andrea Champlin: Remember something happened to his first one.
Mildred: Right. It burned down.
David Adams: Our current agreement with the Park Service, or the Department of the Interior, prohibits renting.
Mildred: But this was in…
David: Prior to the acquisition by the Park.

Mildred: Hawthorne's son…oh, what was his name? He was the orchestra conductor for the Provincetown Symphony and also the Toledo Symphony. He and his wife came out and rented the house for the summer, for several weeks. It was just because they asked us and we said, "Sure."

Andrea: It was never the intent to rent these houses out.

Mildred: Because of our jobs – we were teachers so we were able to take summers off. We came out and spent the summer.

Maia Champlin Peck: We also found that it was harder to rent the houses out here than it is a house in suburbia, because you have to know the water system and you have to know the gas system.

Nat Champlin: I decided to never, ever rent.

David: I would never ever trust anyone but my son, Tom, or John, or Sally.

In this case, people whom the shack heads knew were allowed to stay in the shacks. But they did not feel comfortable with strangers renting their shacks.

Nonprofit Organizations: Innovative Social Forms

As described in the history chapter, nonprofit organizational involvement with dune shacks developed in response to National Park Service activities, rather than being an historic organizational form in dune shack society. Currently, two nonprofit organizations (Peaked Hill Trust and Provincetown Community Compact) were caretakers of four shacks. Peaked Hill Trust came into existence when dune shack users, shocked by the bulldozing of Charlie Schmid's shack, organized with the goal of protecting the remaining shacks and their uses against removal by the National Park Service. The core of that organization (leaders such as Bill and Hatty Fitts, Barbara Mayo, Julie Schecter, and Josephine Del Deo) has been drawn primarily from the network of friends of dune shack resident families, that is, friends who have developed a passion for dune shacks as invited users, but who personally never owned a shack. Provincetown Community Compact, the other nonprofit involved with the shacks, developed not with dune shacks in mind, but to promote the art, natural environment, and culture of Provincetown. The organization became a caretaker of a dune shack by winning a solicitation from the National Park Service to manage the Cohen shack. Its core leaders, Jay Critchley and Tom Boland, were year-round residents of Provincetown.

According to Hatty Fitts, the core organization of Peaked Hill Trust was composed of a twelve-member board and two additional caretakers. This core did most of the planning, scheduling, and week-to-week caretaking of the shacks, such as cleaning, hauling out trash, driving people to and from the shacks during the main season, and so forth. She differentiated this core from a pool of volunteer workers (about 100 members) who intermittently helped during two scheduled workweeks. As described by Hatty Fitts, the remainder of the organization's membership (about 600 members) might be interested in the shacks, but did not normally contribute labor to the program:

The board [of Peaked Hill Trust] is an active working board. It's not just people who sit around and make policy. They actually are involved in the maintenance of the shacks, getting the supplies, and all of that. Of the twelve board members, ten of them are caretakers. The caretakers are the real core. Then we have a couple of others who aren't on the board who are also caretakers. For the most part, these people have been involved with Peaked Hill Trust since the beginning. They're the ones who are willing to put in days at a time to do whatever is necessary to keep things going.

Then there is a secondary group of members who just like to be out there pounding nails, sweeping, cleaning, whatever. They aren't necessarily the same group all the time. It depends on who is doing what at the time we have major work.

In the spring, about the first week of May, we have what we call "workweek." That's when we not only open the shacks, but we do whatever major repairing of whatever happened during the winter that requires repairs, like replacing rot, repainting outhouses, that sort of thing. That stuff is done during the workweek. The other work session that we have is "closing," which is the end of October or beginning of November. That's just a matter of packing things away so the mice don't have something to play with, and closing things up. We don't do a whole lot of repairs at that time, so we don't need as big a group. Those are the two main periods.

The workers are the board, the few others who are also caretakers, and this secondary ring. The secondary ring can be made of up most anybody. There are a few who are regulars, but every year it's a different group that comes to help, depending on what their schedules are. We have 700 members. I would say of the 400 members that I consider are truly interested and involved in the dunes, probably 100 of those are the pool that would be drawn from [for workweek]. Most of the active workers are local or from the area, Boston south to Connecticut.

Though Peaked Hill Trust and the Provincetown Community Compact are recent innovative social forms in dune shack society, their caretaking activities find roots in the traditional culture. As stated above, the core of the organizations are drawn primarily from the network of families and friends with long-term ties to the dune shacks. To a large extent, the boards of the two organizations perform functions that parallel those of the extended family groups at the cores of the family-based shacks. By contrast, the pool of applicants for shack time outside the cores represents a relatively new, atypical segment of dune shack society. Many applicants for stays at a shack have had no previous ties to dune shack culture. This growing group of first-time shack users, selected by a random draw, was the least traditional aspect of the nonprofit programs, as described in the next section.

Lottery and Artist-in-Residence Winners: Non-Traditional Social Forms

The newest type of dune shack user, existing since about the late 1980s, consisted of winners of lotteries and artist-in-residence awards. Generally, these were short-term users participating in one of several programs offered by nonprofit organizations caring for certain dune shacks. As described by Schuster, this was not a traditional use of dune shacks. This type of user developed because of programs run by nonprofit organizations at the request of the Seashore. The uses emerged out of formal arrangements between the nonprofit organizations and the Seashore.

The Provincetown Community Compact ran both an artist-in-residence program and an open lottery program for the C-Scape shack (the former Jean Cohen shack). The Seashore specifically requested an artist-in-residence program in their solicitation for proposals for managing the shack. The C-Scape shack is the most westerly shack in the western group, a thirty-minute walk from the Province Lands Visitor Center. As part of the current artist-in-residence program at the C-Scape shack, the Seashore offered to visitors twice-weekly guided walking tours of the dunes that stopped at the shack. The Seashore's guide interpreted the dunes and shack for tourists, who then had an opportunity to meet the artist-in-residence and learn of his or her work at the shack.

Jay Critchley and Tom Boland, heads of the Provincetown Community Compact, described to me how the lottery and artist-in-residence programs operated for the C-Scape shack:

> Critchley: Provincetown Community Compact was formed in 1993. It's a nonprofit organization whose mission is to enhance the arts, environment, and culture of Provincetown. The organization was formed as a vehicle for artists and other grassroots projects to have a fiscal agent for funds to develop projects for artists, or other community projects, one of them being the Provincetown Swim for Life. One of the projects we took on is the C-Scape dune shack lease with the National Park Service.

> Boland: The Park Service, from what I was led to understand, has a goal of trying to do artists-in-residency programs in all of their parks where they have historic buildings. They realized that they needed to do one here. This shack was a logical fit for a number of reasons – its size, its accessibility, its being not quite as rustic as some of the other shacks. They put that out to bid, for people to come back with creative proposals.

> Under our lease agreement we do three, three-week "artist residencies" in the summer. The artists are selected by jury review. The jury is composed of a number of artists in the community and usually a couple of people representing the Park Service. It's an arduous process of reviewing the applications.

> The rest of the year, meaning spring and fall, we do "community residencies." Those are one-week stays. The people who stay then are selected by a week-by-week lottery. People apply. They tell us what weeks they are interested in, and then we pull their names out of a hat.

> Critchley: We really see this program as a community program. To us, it's the community getting time in the shack. So anyone can apply. We have a sliding fee scale. For the nine-week artists-in-residency program, one of the three-week periods we offer a $500 fellowship and free rent for three weeks. In the other two three-week residencies, the artists pay from $100 to $400 a week, which generally turns up being $100 a week. It's a sliding scale as well for the community residencies. Provincetown is such an artist's community that nobody pays artists for anything. Artists are always asked to donate pieces of work, of art, to all these auctions and benefits. This goes on and on and on. So we felt very strongly that at least one artist should get paid, which is another way that this program connects to the town

So the C-Scape shack operated two types of shack occupancies. One was the artist-in-residence program, a three-week stay. Awards were made through a juried selection of proposals. The other was an open lottery to anyone for one-week stays during spring and fall. Rents were charged, except for one artist-in-residence who received a fellowship. In addition, the shack was sometimes opened during winter. As previously described, a third type of user of the C-Scape shack consisted of people receiving shack time for labor.

According to Hatty Fitts, there were about 60 slots of time available under the lottery programs and artist-in-residence program operated by Peaked Hill Trust, after the time for caretakers and shack heads were filled. Julie Schecter described this program, beginning with a description of the shacks currently being cared for by the organization:

> There are four shacks. Hazel's two shacks were the first two – Thalassa and Euphoria. The third one we manage is Boris's shack, known as the Margo-Gelb. It was the same kind of

situation as Hazel's. He wasn't really using it anymore and gave us permission. We were doing all of this by way of a demonstration, to show that the shacks could be used in a fair and low impact way. Then Zara Ofsevit's shack [the Jackson shack] burned down. She had a timed lease which included the ability to rebuild should anything happen. So Peaked Hill did the fundraising and the physical labor of rebuilding. The agreement was that she would get to use it for a specific time every year, and that the Peaked Hill would get to use it for the rest of the time.

The Margo-Gelb shack, Boris's, is managed somewhat differently from the others. It has the artist-in-residency program that's part of the Park Service. In that program, artists apply for two-week stints in the shack. Five or six artists each year are given some time out there. We have had artists who have gone out there and leave after a week, because it's not their style, or they had other pressing engagements, but those are two-week slots that are saved for the artists. Peaked Hill uses the rest of the time for its members.

In the other shacks, a different guest will go out to each shack each week. Those individuals are sort of subject to a series of hurdles if you will. If you want to have shack time, if you want to have a week in a shack, you have to be a member of the organization as of January 1 of that year. You must respond to a letter asking you if you want to use time in the shack this year and tell us when and where you would like to be. All the names go into a metaphorical hat. We stir the names around and pull them out. Random number generator now, but originally it was a hat. If you asked for time last year and you didn't get it, and you ask again this year, your name goes in twice. If you've been denied two years in a row and you ask for the third year, your name goes in three times. So over time, you get time in the shack. We have a waiting list of about five years. There's about five times the applicants as space available. There are only about twice as many applicants for the space available for the artists' shack time.

Peaked Hill was designed from the beginning to allow for public access even though it's a membership organization. Anybody can join. The membership part of it has to do with trying to organize, not just putting "Shack For Rent" out on the street, trying to allow people to plan their time, and allowing Peaked Hill to plan things. You're just not inviting anybody in, but anybody can come in. The artist-in-residence program is for artists only, but it is for any artist.

As described by Schecter, the Peaked Hill Trust programs resembled that of the C-Scape shack, with two tracks – one designed to accommodate artists-in-residence and one designed for anyone. A jury selected the artists. A randomized process selected applicants from the general public. With the Peaked Hill Trust program, a person joined the organization to participate, requiring a $25 membership fee.

Peaked Hill Trust counted the numbers and origins of persons awarded time at their shacks during a ten-year period, from 1986-1995. This summary counted 310 visitor awards. The greatest number of visitor awards (115, or 37 percent) was to people who gave Cape Cod resident addresses, including Provincetown (69 awards with 87 people), Truro-North Truro-Wellfleet (26 awards with 29 people), and other cape communities (20 awards with 22 people). In addition, visitor awards drew from Other Massachusetts (85), New York (35), Connecticut (16), and 55 from 33 other states. These numbers did not include shack time given to caretakers.

As stated above, the uses of lotteries and juries for placing occupants in dune shacks are not traditional practices. These selection mechanisms have no historic counterparts prior to the formation of the Seashore. The impersonal lotteries and juries contrast with two underlying features of dune shack society: the personal nature of extended families and friendship networks, and the knowledge base of long-term residents versed in dune shack history and culture. The lotteries and juries potentially placed in the nonprofit's shacks a different type of user – short-term occupants drawn from the general public with no historic ties to, or personal knowledge of, the local culture and its traditions. The impersonal selection processes represented a potentially significant impact on shack society if the general public substantially replaced or eliminated long-term residents of dune shacks, the core set of people who were the tradition bearers of the dune shack culture. Without tradition bearers to pass on customary patterns at the shacks, dune shack culture would probably cease as a living tradition.

Contrasting Forms

In looking at the organization of dune shack society, I did not observe a number of social forms that were common elsewhere on Cape Cod. Missing on the Backshore were the "summer vacation rental," "timeshare condo," "motel," "hotel," "hostel," and "government bunkhouse." These types of lodging were not part of dune shack patterns. The summer vacation rental was a prevalent form elsewhere on Cape Cod. It was short-term lodging for visitors paying money rent, operated as a business within the cape's large summer tourism industry. As stated above, certain dune shacks occasionally rented historically, typically as extensions of personal relationships between shack users involving informal exchanges of money, labor, or services. But this traditional dune shack pattern was substantially different from the impersonal rental markets serving the crowds of summer visitors to Cape Cod, where today's lodging was advertised to the general public on the Internet.

I learned of "timeshare condos" at a Laundromat in Orleans, chatting with a middle-aged social worker from Washington D.C. who regularly vacationed on Cape Cod. That afternoon the Laundromat was filled with vacationers doing wash, including herself. She said that this year she was renting a timeshare condo at $500 per week. The building was originally run as a motel. The former motel owners recently had converted it into a condominium, offering tiny efficiency units that sold for about $130,000 each. To new condo owners who purchased a unit, the former motel owners sold additional services, such as maintenance and reservation booking to fill dead time. For them it was like "having your cake and eating it too," she said. The social worker was exploring the pros and cons of investing in such an arrangement, to see if she could have a vacation place while making a profit renting to others. In the meantime, she was "scoping out" new places for renting next year. She was searching for a cottage near a beach "without a tall seawall to climb over." She said the competition for good spots was "fierce." The types of social arrangements described by this vacationer were not found on the Backshore. As discussed in the historic chapter, grassroots political action at Provincetown during 1959-61 was directed toward protecting the Province Lands and Backshore dunes against this kind of real estate development.

Another institutional arrangement not found on the Backshore was the "government bunkhouse." Historically, for brief periods during the Second World War, certain dune shacks were used to house military personnel. The Braaten shack (now the Schuster shack) was used to support torpedo testing by the Navy, according to Lawrence Schuster, the current resident. Other shacks, such as the Avila shack (now the Champlin shack), may also have briefly assisted the military (handing out food to beach patrols). With these types of exceptions, the dune shacks have not been institutional housing for government workers. During this project, I learned firsthand of government housing by staying at a cottage in Eastham on the shore of Nauset Marsh owned by the National Park Service. It had been a family-owned cottage. The Seashore purchased it to house seasonal researchers on government-related projects. From its picture windows, I looked directly across the beautiful tidal marsh to the sand spit where Henry Beston's famous Outermost House once sat. During my month-long occupancy, I shared the bunkhouse, off-and-on, with several other research groups, including university researchers conducting a public-opinion survey on hunting on the Seashore, a biological team sampling fish in nearby wetlands, and a geomorphologist who turned out to have once owned a dune shack outside Provincetown. The shuffling of people in and out of the house was efficiently scheduled from the Seashore's headquarters. This type of social arrangement is common within institutional cultures that support seasonal research. As stated above, the social form was foreign to dune shack traditions.

www.ingramcontent.com/pod-product-compliance
Lightning Source LLC
Chambersburg PA
CBHW081850170526
45167CB00007B/2962